DIRTY GENES

DIRTY GENES

A Breakthrough Program to Treat the Root
Cause of Illness and Optimize Your Health

DR. BEN LYNCH

HarperOne
An Imprint of HarperCollinsPublishers

HarperOne

DIRTY GENES. Copyright © 2018 by Dr. Ben Lynch LLC. All rights reserved. Printed in the United States of America. No part of this book may be used or reproduced in any manner whatsoever without written permission except in the case of brief quotations embodied in critical articles and reviews. For information, address HarperCollins Publishers, 195 Broadway, New York, NY 10007.

HarperCollins books may be purchased for educational, business, or sales promotional use. For information, please email the Special Markets Department at SPsales@harpercollins.com.

FIRST EDITION

Designed by Ad Librum

Library of Congress Cataloging-in-Publication Data

Names: Lynch, Ben, 1974– author.
Title: Dirty genes : a breakthrough program to treat the root cause of
    illness and optimize your health / Dr. Ben Lynch.
Description: First edition. | New York, NY : HarperOne, [2018]
Identifiers: LCCN 2017027349 (print) | LCCN 2017037582 (ebook) | ISBN
    9780062698209 (E-book) | ISBN 9780062698148 (hardback) | ISBN
    9780062800855 (audio)
Subjects: LCSH: Genetic regulation. | DNA—Methylation. | Medical genetics. |
    Health—Nutritional aspects. | BISAC: HEALTH & FITNESS / Nutrition. |
    HEALTH & FITNESS / Diseases / Immune System. | HEALTH & FITNESS / Healthy
    Living.
Classification: LCC QH450 (ebook) | LCC QH450 .L96 2018 (print) | DDC
    572.8/65—dc23
LC record available at https://lccn.loc.gov/2017027349

18 19 20 21 22   LSC   10 9 8 7 6 5 4 3 2 1

This book is dedicated to the late Rachel Kranz. She brought *Dirty Genes* to life, although she did not live to see it published. She worked tirelessly to better the health of millions through books. Without her tremendous ability to organize, create, write, strategize, and collaborate, there would be far fewer amazing health books available today. She gave it her all. I know of no one who has impacted so many lives so deeply, so selflessly, without any need for public acknowledgment. Without Rachel, this book could not have been written. She inspired me, coached me, pushed me, challenged me, and encouraged me each step of the way.

CONTENTS

Part III Your Clean Genes Protocol

Your Genes Are Not Your Destiny!

It was just an ordinary day in 2007. I had half an hour to spare and decided to check out a program playing on the PBS show *Nova*, "A Tale of Two Mice."

The program introduced us to two mice that were genetically identical—but looked completely different. Both were from a strain that had a strong genetic potential for obesity, cardiovascular disease, and cancer. Yet one of the mice was lean and healthy, while the other was massively overweight and vulnerable to disease. Although each had the genetic *potential* for major illness and excess weight, only one of them was actually unhealthy.

As I watched in astonishment, the researcher explained the "x factor"—the mysterious, powerful reason behind our ability to manipulate our genetic inheritance and create health rather than illness. The secret was *methylation*, a biochemical process that takes place within your body. By methylating certain genes, you can *turn off* your genetic tendency to obesity and disease.

And how had that amazing feat been accomplished in the mice presented on PBS? In that experiment, through diet alone. While the experimental mice were still in the womb, researchers had given some of the mothers *methyl donors*—nutrients that support the methylation process—while the control group received none. The right diet had turned off the mice's "dirty genes" and reshaped their genetic destiny.

This process of turning genes on and off is known as *epigenetics.* What I have learned since that momentous day in 2007 is that we can transform our genetic destiny through a combination of diet, supplements, sleep, stress relief, and reduced exposure to environmental toxins (the toxins in our food, water, air, and products). With the right tools, we can transcend our inherited tendencies to disease—including anxiety, attention-deficit/hyperactivity disorder (ADHD), birth defects, cancer, dementia, depression, heart disease, insomnia, and obesity—to create new and healthy lives.

I still remember how amazed I was as the program drew to a close. I slammed my hand down on my desk. "That's *it!*" I cried out. "*That's* what I want to do!"

From that point on, I was obsessed. Contrary to what so many scientists and doctors believe, our genetic destiny isn't fixed. It can be edited, rewritten, changed. We just need to know how.

And so it became my mission to identify our dirty genes and develop the protocol we need to scrub them clean, replacing disease with health and enabling us all to reach our genetic potential. I'm happy to tell you that after a decade of research, study, and successful treatment of clients around the world, I have developed and refined the Clean Genes Protocol, a program to optimize your health—and your life.

The Power of Epigenetics

I've always been fascinated by the ways our bodies *want* to be healthy, and I've spent most of my life learning how to help them get there. As an undergraduate, I studied cell and molecular biology. I then became a naturopathic physician—a science-based practitioner who relies on natural methods to restore balance and optimize health. As I worked with patients, I realized that I also needed to become a specialist in environmental medicine, discovering both how the chemicals in our environment undermine our health and what we can do to detoxify our bodies.

What made all my diverse studies come together was the field of epigenetics: the many, many factors that can influence how our genes are expressed. I had always understood how powerful genes can be. But *wow*, was I thrilled to discover that we don't have to bow down and submit to our DNA. Instead, we can work *with* our genes to create optimal health—if only we know how.

One of the most important pieces of the genetic puzzle is a type of variation known as a SNP (pronounced "snip"), which is short for *single-nucleotide polymorphism*. So far, roughly ten million SNPs have been identified in the human genome, with each of us having over a million.

Most of those SNPs don't seem to affect us very much. Yes, they represent a slight variation or abnormality in various genes, but so far as we know, those variations don't seem to make much difference in the way our bodies function.

Some SNPs, however, can make a huge difference in our health—and in our personality, as well. For example, SNPs in the MTHFR gene can create a whole host of health problems—everything from irritability and obsessiveness to birth defects and cancer. (Note that I said *can*. They don't *have to*—that's what this book is all about!) SNPs in the COMT gene can lead to workaholism, sleep issues, PMS, problems with menopause, and again, cancer, along with boundless energy, enthusiasm, and good spirits. (Yes, many SNPs have an upside as well as a downside!)

Health issues that had puzzled my clients for years suddenly made sense when they discovered through our work together that their SNPs had at least partly created those issues. Problems that had seemed overwhelming—even dooming—became manageable as clients learned that they could use diet and lifestyle to reshape their genes' behavior.

I experienced that kind of *aha* moment myself when I discovered that I have at least three significant SNPs. I finally understood something about why I'm so focused and determined—some might say obsessive! I also saw why I can suddenly become irritable at a moment's notice, and why I react so intensely to certain chemicals and fumes. It was a relief to have this new understanding: things made sense to me in a way they never had before, plus I had some new solutions I could rely on. As you read this book, you'll have the chance to make the same kind of exciting discoveries about yourself.

Most important, learning about my SNPs empowered me to take charge of my health. Finally, I could support my body and brain with the diet and lifestyle that they needed. For the first time in my life, I felt like I was working at the top of my potential.

I wanted my patients to have the same experience—heck, I wanted *everybody* to hit that high. So I began to develop a program for cleaning up our "dirty genes": what we should eat, which supplements could help, and how to create a "clean genes" lifestyle. I wanted us to be like that epigenetically supported mouse, glowing and healthy, no matter which genes life has dealt us. I knew that if I could just delve deeply enough, I'd find the answers.

And now, ten years later, I'm proud to say that I have. Oh, there's still a ton more I have to learn—the whole field of epigenetics is just getting started, and we're making new discoveries every day. I spend a big portion of every week doing my own research, and another big chunk of time reading the studies that my colleagues turn out. It's more than one person could ever keep up with—and that's the *good* news. In another ten years, I'm fully confident, we're going to have the power to take charge of our health in ways we can't even imagine.

Still, you want to be healthy *today,* not in a year or two or ten. So besides my research and study, I've worked with thousands of clients and hundreds of doctors, learning how all that arcane science can be translated into a practical, accessible program that anybody—no matter how busy—can do. I speak at conferences, publish videos, and maintain a blog, reaching physicians, health professionals, scientists, and laypeople.

I'm proud to say I've become the go-to guy on epigenetics even for such nationally recognized doctors and bestselling authors as Chris Kresser, Izabella Wentz, Alan Christianson, Peter Osborne, and Kelly Brogan.

To be honest, I wish I *wasn't* the primary source of information—I wish what I knew was commonplace and widespread, the basic approach available in every doctor's office. That's become my new goal, and it's why I've written this book: to teach every single interested person how to clean up their genes and achieve a whole new level of health.

Are Your Dirty Genes Creating These Problems for You?

Brain and Mood Issues
- ADD/ADHD
- Anxiety
- Brain fog
- Depression
- Fatigue
- Insomnia and sleep problems
- Irritability
- Memory problems

Cancer
- Breast cancer
- Ovarian cancer
- Stomach cancer

Cardiovascular Issues
- Atherosclerosis
- Heart disease
- Hypertension
- Stroke
- Triglyceride elevation

Female Hormone Issues
- Menopause difficulties
- Menstrual challenges—cramps, excessive bleeding, mood and cognition difficulties
- Menstrual migraines
- PMS

Fertility and Pregnancy Issues
- Difficulty getting pregnant
- Difficulty carrying to term
- Increased risk of birth defects

Gland and Organ Issues
- Fatty liver and other liver dysfunction
- Gallstones

- Small intestine bacterial overgrowth (SIBO)
- Thyroid dysfunction

Metabolism Issues
- Food cravings, especially for sweets and carbs
- Obesity and weight gain

Are Your Dirty Genes Making You Sick?

You've probably heard that your genes affect how healthy you can be. Almost certainly, your doctor has told you that because of the conditions that run in your family, you may be vulnerable to heart disease, depression, anxiety, and/or other disorders.

Most of the time, this news makes people feel discouraged. "I'm scared," they tell me. "My genes are a mess. I've just gotta make the best of it."

No way!

After years of research in the new science of gene abnormalities, and having successfully treated thousands of clients, including my family and myself, I'm offering you an exciting new approach: a proven method to clean up your genetic limitations and create a healthier, more vibrant you.

So let me say it loud and clear: *Your genes are not your destiny!*

But perhaps you've been taught that they are. Perhaps, like most people, you've been told that you inherited a "master plan"—characteristics that are written in stone from the moment of your conception until the day you die. In this view, your genes are a stern committee of judges handing down a life sentence.

"Hmm," your genes seem to say. "Let's give this woman depression, which she gets from her mother. And let's throw in some heart disease, which runs on her father's side. How about adding in a shy, anxious personality, which she gets from her grandmother? We're almost done, but let's throw in one more ingredient—a mild case of ADHD, nothing clinical. But, like two of her uncles, she's always going to have a hard time focusing. There—all done! Good luck, lady! Enjoy the destiny that we've written for you, because there is absolutely nothing you can do to change it!"

Pretty bleak, right? Luckily, it's *wrong*. Far from being written in stone, your genetic destiny is more like a document written in the Cloud—you get to edit and revise it, every moment of your life. Every time you drink a soda, barrel through on four hours of sleep, use a shampoo loaded with industrial chemicals, or hit a stress bomb at your job, you're putting the negative part of the document in giant type. And every time you eat some organic leafy greens, get a good night's sleep, use a chemical-free shampoo, laugh with friends, or do some yoga, you're enlarging the positive part of the document while reducing the negative part to a font so small, it might as well not be there at all.

Your genes don't lay down the law; they negotiate with you. They don't even speak with a single voice. They're a committee, and sometimes they disagree with each other.

Some of the folks on that committee are harsh. They're constantly shouting, "Heart disease!" or "Depression!" or "Crippling lack of confidence!" And if you don't know the right way to work with them, those loud, harsh voices might rule the day.

But—and here's how this book is going to change your life—if you *do* know how to work with your "gene committee," you can produce a *much* better outcome. You can get those loud negative voices to tone it down or even shut up completely. At the same time, you can turn up the volume on the voices that are saying, "Balanced mood!" "Healthy heart!" and "Self-confidence!"

So get ready to clean your dirty genes, folks, because that's exactly what you're going to do. In this book, you'll find out how to make the most of your genetic inheritance, now and for the rest of your life.

Are Your Genes Dirty? Some Common Symptoms

- Aching joints and/or muscles
- Acid reflux/heartburn
- Acne
- Allergic reactions
- Anger and aggression
- Anxiety

- Attention issues
- Blood sugar spikes and crashes
- Brain fog
- Cold hands and feet
- Constipation
- Cravings, especially for carbs and sugar
- Depression
- Diarrhea
- Edginess
- Fatigue
- Fibromyalgia
- Food intolerance
- Gallstones
- Gas and bloating
- Headache/migraine
- Heart racing
- Indigestion
- Insomnia
- Irritability
- Itchy skin
- Menopause/perimenopause symptoms
- Mood swings
- Nosebleeds
- Obesity/weight gain
- Obsessiveness
- Overreactive startle reflex
- PMS / difficult periods
- Polycystic ovarian syndrome (PCOS)
- Rosacea
- Runny nose / congestion
- Sweating
- Unexplained symptoms—just "not feeling right"
- Workaholism

What Your Doctor Won't Tell You—But I Will

If you've had trouble with any of the symptoms I've just listed, your doctor might have told you that you're not *really* sick. Or maybe you've been offered drugs to medicate the symptoms—antibiotics, painkillers, antacids, antidepressants, anti-anxiety medications—without much attention to the underlying issues that produce those symptoms.

Or perhaps you've been one of the lucky ones. Perhaps you've found a naturopathic physician, functional/integrative MD, osteopath, nurse practitioner, nutritionist, chiropractor, or other health professional who has helped you restore health and wellness through diet, lifestyle, and other natural means. Even so, your treatment is incomplete if you haven't learned about dirty genes, the root cause of many of the conditions you're struggling with.

That's because epigenetics—modifying genetic expression to improve your life and health—is a cutting-edge aspect of medicine that most practitioners don't understand. I'm one of the few people who have figured out how to translate genetic research into concrete actions that will improve your health, which is why so many leading health-care specialists come to me for training and advice. That's why I spend so much of my time lecturing and consulting with physicians and providers—conventional, natural, and otherwise—as well as reading other people's studies, doing my own research, and helping others regain their health.

As a result, the suggestions in this book are based on the very latest scientific material. Most health-care providers simply aren't aware of this information, though I'd be willing to bet that in a few years, programs like the one in this book will be widespread, even standard.

Meanwhile, congratulations! Reading this book puts you well ahead of the curve. Once you understand what makes your genes dirty and how to clean them up, you'll feel better than you ever thought you could.

What Kind of Doctor Am I?

I am a naturopathic physician: an ND, or naturopathic doctor. *Naturopathy* is a science-based system that uses natural methods to restore health and wellness, with a focus on treating the underlying causes rather than the symptoms and relying on natural means: diet, lifestyle, herbs, supplements, avoidance of chemicals, support for detoxification, and reduction and/or relief of stress. A growing number of medical doctors, nurses, and other health-care providers practice a similar approach known as *functional/integrative medicine,* which also focuses on natural means to treat root causes.

My Own Dirty Genes Success Story

My personal struggle with dirty genes has been long and confusing. Even as a kid, I thought there must have been *something* behind my challenges, but for the life of me, I couldn't figure out what.

Intense and superfocused, I could also fly off the handle at a moment's notice, becoming irritable and frustrated without warning. I seemed to feel things more than most people I knew, to be less patient but also more determined. In addition, I had horrific stomach pains more often than I'd like to remember. Also lifelong low white blood cells until just a few years ago, as well as intense sensitivity to chemicals and cigarette smoke.

Later I would realize that these traits correlate strongly to specific SNPs. I would figure out that there were foods I could eat—and foods I could avoid—that helped me emphasize the positive side of these traits while reducing or even eliminating the downside. I also discovered that sleep, stress, and toxic exposure played a huge role in how my SNPs affected me—that diet and lifestyle affected the expression of many of my genes.

Meanwhile, I enjoyed my teenage years growing up on my family's horse ranch in central Oregon, where I learned to work hard, to be self-motivated, and to appreciate nature and the cycle of life. In 1995, I was a restless student and athlete at the University of Washington who got bitten by the travel bug—hard. I took a one-year leave to backpack through the South Pacific and Southeast Asia.

What an amazing trip! I did everything from living a bare-bones subsistence life on the remote island of Somosomo to working as a jack-aroo on a 1.5 million–acre ranch in the Australian outback. I somehow made it to India, where I volunteered with Mother Teresa and her sisters.

And I became deathly ill. You know it's bad when stuff comes out both ends at the same time repeatedly—especially when you're just standing next to a poor unsuspecting camel in the middle of the street.

I'm still not sure what I had. Far from any standard medical center, I didn't have any choice but to settle for whatever type of diagnosis and treatment was on offer. My single option was Ayurveda, the ancient Indian medical tradition that focuses on food, herbs, and lifestyle rather than on factory-made pharmaceuticals. When it cured me, I saw for myself the power of natural healing.

Eventually, I made my way to Bastyr University, the leading educational institution for naturopathic medicine. But not before a checkered career that included earning a B.S. in cell and molecular biology from the University of Washington, rowing on the Huskies crew team, traveling to more than forty countries on a shoestring budget, founding a high-end landscape construction firm, and summiting both Mount Rainier and Mount Baker. While studying at Bastyr and working with a wide variety of naturopathic health-care providers, my life got even busier. I got married, had three sons, traveled frequently, and started a new business, selling supplements and other products to support people's health.

Ironically, even as I learned how to make other people healthy, I worried about myself. After all, there was a lot of cancer, alcoholism, and stroke in my family. Was that my destiny, too? I knew that diet and lifestyle could play a huge role in making a person healthy. But I couldn't stop thinking about the genetic piece of the puzzle.

In 2005, I was working with a leading physician who specialized in environmental medicine. He had developed a powerful protocol for people who were succumbing to the toxic effects of heavy metals and industrial chemicals. Most patients did very well on his protocol. But some had no improvement—and some of them got worse.

"Do you think this is about genetics?" I asked him. "Do some people have genes that make it difficult for them to clear out the chemicals?"

It was an intriguing question—and my mentor didn't have the answer.

But when I saw "A Tale of Two Mice" in 2007, I realized that genes and environment did indeed work together to shape our health. The key was to figure out how our genes got dirty in the first place—and how to clean them up.

Two years later, a colleague asked me what natural methods could help a patient struggling with bipolar disorder. I started rattling off the usual answers, but I stopped myself. I had been out of school for a couple of years—maybe I had missed some new research.

After three hours at my computer, I was amazed. I had learned that there was a connection between bipolar disorder and SNPs in the MTHFR gene, which was exciting enough. But digging deeper revealed that MTHFR SNPs were also implicated in many other major health issues, including anxiety, stroke, heart attack, recurrent miscarriage, depression, Alzheimer's disease, and cancer.

How had I not known about something this significant? I had to look further. The more research I did, the more important MTHFR came to seem.

Eventually, I had myself and my family tested. I was horrified to find not only that I had a number of MTHFR SNPs, but that two of my sons did, too. I felt as though I had been hit over the head and punched in the gut simultaneously. How could I protect us?

So I got to work. I began learning still more about MTHFR—and in the process, I discovered many other dirty genes. For the next ten years, I divided my time between research, study, and trial and error, trying to help my clients clean up their dirty genes. Finally, I put together a protocol for the overall "Soak and Scrub" (presented in chapter 12), and I learned how to do a more focused "Spot Cleaning" (chapter 15) to address specific genetic issues. What a relief! I could be healthy. My sons could be healthy. And so could the hundreds of thousands of people whom I reached directly as clients or indirectly as participants in the workshops I gave for physicians, nutritionists, and other health professionals.

This book is the final step—a way to share my discoveries with you and your family. Whether or not you've had genetic testing, whether or not you've ever even considered the role of genes in your health, this book can empower you to optimize your health—and your life.

Your Dynamic Genes

Remember, every moment of every day, your genes are working on that document about your health. They can write it in a way you like or a way you don't like—but they're always writing. And whether you know it or not, so are you.

For example, your genes keep telling your body, "Rebuild your skin!" As you know if you exfoliate, your skin is constantly dying and being replaced. So every moment of every day, your genes are adding to the document, telling your body to get on with the repair.

What kind of document do you think they write if you eat a high-sugar diet, skimp on sleep, or stress yourself out for days on end? Hmm, maybe something like: *Please give this woman dull, lackluster skin with plenty of acne and maybe a touch of rosacea.* On the other hand, provide your genes with healthy fats, plenty of sleep, and time to chill, and you're going to see a different document: *This one gets healthy, glowing skin that makes her look ten years younger!* Your genes won't stop writing until the day you die. But *what* they write is up to *you*.

Likewise, your genes are constantly producing documents about your gut lining, which is repaired and rebuilt every seven days. If you eat right and live right, you're going to get a great document: *Keep that guy's gut strong and healthy!* If you mess up your genes with poor diet and lifestyle, your document probably says something like: *Since this man is giving me so much extra work to do, I can't focus on repairing his gut lining. He's also not providing me the tools I need. So give this man a weak gut lining—the kind that lets food leak through.* Watch out for all the weight gain, immune issues, and other problems that are likely to follow!

Now here's my personal favorite: the memo about your mind. *Those* instructions involve *neurotransmitters*—biochemicals such as serotonin, dopamine, and norepinephrine that govern your thoughts, moods, and emotions. Your brain runs on thousands of biochemical reactions, and there are countless ways the process can go wrong. Your goal is to give your genes whatever they need to produce an uplifting memo: *Keep this person sharp, focused, calm, and full of energy during the day, and relaxed, calm, and ready to sleep at night.* The memo you *don't* want mentions forgetfulness, depression, anxiety, irritability, insomnia, addictions, and brain fog.

So yes, your genes write your life memo. But *what* they write is largely up to *you.*

Sound good? Then let's get started. Turn the page to read about some of my patients' most dramatic and inspiring success stories—and know that you can enjoy that success too, as soon as you clean your genes.

PART I

CAN YOU CONTROL YOUR GENES?

1

Cleaning Up
Your Dirty Genes

When Keri and I began our session, she was distraught. She held a wad of tissues in one hand, and with the other she was continually wiping her runny nose or dabbing at her streaming eyes. Her skin was red and scaly. Her hair was limp and stringy. Almost before I could introduce myself, she burst out, "I'm such a mess!"

As we talked, I could see that Keri already understood quite a bit about her situation. She had already figured out what was making her sick: chemicals. "Just a whiff of paint starts me wheezing," she told me, between efforts at repair work with the tissues. "Every time I clean the kitchen floor, my eyes tear up. I can't even find a shampoo or a bar of hand soap that doesn't make me break out. I've tried to clear all that stuff out of my house—but every day, it seems like I develop a reaction to something new. I feel like I'm going crazy—but I'm not, am I?"

No, I reassured her. From her symptoms, I was willing to bet that Keri had at least one dirty gene. More specifically, I suspected one or more SNPs in her GST or GPX. Those are the genes that help us use *glutathione*, a key detox agent that our body produces. Without gluta-thione, we have a heck of a time ridding our body of toxins. And in our modern world, we're surrounded by toxins all the time. Industrial chemicals and heavy metals are in our air, our water, our shampoo, our face cream, our food, our dish soap, our laundry detergent—the list goes on and on. And on.

Yes, your genes will thank you for buying organic and using only green products—that's a great start. But your body also has to filter out the toxins that you just can't avoid. How about the daily eleven thousand liters of air you breathe, the eight cups of water you drink, and the four pounds of food you eat? These are all infused with at least some of the 129 million industrial chemicals currently registered. And filtering those chemicals out of your body is nearly impossible when your GST/GPX gene is dirty. (Because it's hard to tell, without genetic testing, which of these two closely related genes is the culprit, I often refer to them jointly as the GST/GPX gene. If you have other dirty genes, the whole process is even harder. To get rid of her symptoms, Keri was going to have to go beyond buying organic, and clean up her genes.

Jamal was nervous, and rightfully so. He came to me because both his grandfather and his uncle had passed away in their fifties from a heart attack. Now Jamal's fifty-six-year-old father was also seeing a doctor for cardiovascular issues.

"I'd like to understand what's going on with my family," Jamal told me. "I feel like I'm facing a death sentence, and I don't want to be next."

No, I assured Jamal, he was most definitely not facing a death sentence. And I was impressed that he was being so proactive in taking charge of his health. Yes, from the number of cardiovascular issues that ran in his family, he very likely had been born with a dirty NOS3, a gene that plays a central role in heart function and circulation. His family history was powerful testimony to the way genetic inheritance can affect health.

Can affect health—but doesn't *have to*. A whole world of nutritional and lifestyle support was waiting for both Jamal and his father, treatment options that went far beyond those provided by their doctor.

"You've taken the first steps—and there's so much you can do," I told him. "You just need the right tools."

Taylor had struggled with depression ever since she could remember. As a child, she had been moody and often inconsolable. As a college student, she now struggled with depression and anxiety.

One of her worst problems, she told me, was the way she froze up whenever she had to present in class or take a test. Material she knew perfectly well when she was relaxed seemed to fly out of her head when she was under pressure.

I recognized Taylor's performance anxiety because I had seen it in so many patients—and in myself. I also recognized the mood swings—those blue days when it seemed like nothing would ever go right again. From her symptoms, I became certain she was dealing with a dirty MTHFR gene.

"If your MTHFR is dirty, it can mess up your mental and physical health in a bunch of ways," I told Taylor. That's because MTHFR is crucial to one of the body's most important biological processes—methylation. As a result, a dirty MTHFR creates not only anxiety and depression, but a whole host of other symptoms, including weight gain, headache, fatigue, and brain fog. Cleaning up a dirty MTHFR is a critical step in balancing your mood, improving your performance, and supporting your health.

At first, Taylor was discouraged to think of having been born with a dirty gene. "So, I'm like, what, a mutant?" she asked me. But when I explained that all of us have from one to several dirty genes just among the seven important ones highlighted here, and that the Clean Genes Protocol could enable her to scrub her key genes clean, she became excited at the prospect of overcoming her depression and anxiety for the first time in her life.

Keri, Jamal, and Taylor were all struggling with dirty genes—the root cause of their health troubles. If you're suffering from any of the symptoms on pages 7 through 8, dirty genes are likely at the root of *your* health problems, too.

How Dirty Genes Mess with Your Health

Most likely, neither you nor your doctor is used to thinking about your genes as an active, dynamic factor affecting your present-day health. Instead, your genes seem like an unchangeable, unavoidable set of hardwired instructions passed on from your parents at the moment of conception.

I want you to shift that mindset. Instead of seeing your genetic inheritance as a fixed set of instructions from the past—instructions written on a stone tablet handed down from the ancestors—I want you to see your genes as active participants in your daily health. Right now, while you're reading this, thousands of genes throughout your body are giving instructions—to your brain, digestive tract, skin, heart, liver, and many other aspects of your anatomy. Those genetic instructions shape every facet of your experience and your health, and your genes are handing them out every single second. With every breath you take, every object you touch, every thought you have, you give your genes instructions—and they *respond*.

Let's say you eat a big lunch—too big, more than your body can handle. Oops! Your genes are overloaded. They stagger under the burden of all that food. They tell your metabolism to slow way down. They have trouble methylating—a key process that facilitates at least two hundred functions in your body, from skin repair, digestion, and detoxification to mood balance and clear thought. Because of the challenge posed by that overly large meal, hundreds of instructions are being given differently— and badly. You might promise yourself to eat light that night to make up for it, and maybe you even will. But that won't prevent the damage you inflicted at lunchtime, when you didn't give your genes the conditions they needed to do their job.

Or let's say you stayed up late last night, playing a video game or answering email or binge-watching your favorite show. Now the alarm is going off and you can barely drag yourself out of bed. "I'll make up for it this weekend," you promise yourself—and maybe you will. Meanwhile, though, your genes are living in the present, and they aren't happy about the lack of sleep. They give instructions that alter your digestion, your mood, your metabolism, and your brain, so that right now—not when you were first born, but *now*—your health shifts and slips and declines a little bit.

Of course, if most of the time you're eating well and sleeping deeply and limiting your toxic exposure and managing your stress, an occasional big meal or late night doesn't make all that much difference. Sure, your genes alter their responses for a little while, but your body is strong and resilient, and it can handle the extra challenge. If one gene staggers, a second one steps up. If that second gene stumbles, a third one takes over. Your body has lots of built-in backups, which is terrific.

However, if you consistently give your genes poor working conditions, they're going to consistently hand out poor instructions. Why? Because each backup gene is going to push on the next backup gene, one after the other after the other, and before you know it, too many of your genes are struggling. Your health will suffer, and in way too many cases your doctor won't be able to do much more than prescribe a few drugs to medicate your symptoms.

I want something better for you—a *lot* better. I want you to give your genes exactly what they need to hand out the instructions for perfect health. I want your first-line genes working optimally as often as possible, putting the least possible strain on your backups. I want all your genes cooperating smoothly to give you glowing skin and a healthy weight and tons of energy and a clear, sharp mind. I want you feeling calm and enthusiastic and ready to go, and I want you sleeping so deeply at night that you wake up each morning feeling terrific. If you want that for yourself, listen up: *The way to get optimal health is by supporting your genes.*

Two Types of Dirty Genes

You have *two* types of dirty genes—both of which can give you a host of symptoms and disorders.

Some Genes Are "Born Dirty"

The scientific name for a born-dirty gene is *genetic polymorphism*, which is a fancy way of saying "genetic variation." As we saw in the introduction, these genes are also called single-nucleotide polymorphisms, or SNPs—pronounced "snips." These dirty genes—and each of us has several—can do a full-scale number on your body and your brain. They help determine whether you're heavy or slim, sluggish or energized, depressed or optimistic, anxious or calm.

We have about twenty thousand genes in our body. There are more than ten million known genetic polymorphisms (SNPs), and one person can have as many as 1.2 million of them. However, only about forty thousand are known to potentially alter your genetic function. In this book, we're going to zero in on the key SNPs in the seven genes most likely to have the biggest impact on your health. I chose these Super Seven because each of them influences hundreds of other genes. If any of these seven genes are dirty, you can be sure they're making your other genes dirty, too.

When my clients first discover that they were born with SNPs, many of them are upset. As Taylor said, "I feel like a mutant." But in fact, we're all mutants—that is, every one of us is loaded with SNPs. It's just part of the magnificent variety of the human race—what enables each of us to be unique.

The good news is that once you know which SNPs you have, your health issues start to make a lot more sense—and your emotional issues do, too. If you suffer from migraines, can't seem to fall asleep at night, or struggle with a hair-trigger temper, SNPs may be at the root of your problem. SNPs also contribute to anxiety, depression, irritability, workaholism, obsessiveness, difficulty paying attention, trouble winding down, and a whole bunch of other things that you might never have realized had a genetic and biochemical basis. SNPs also contribute to various strengths, such as boundless energy, good spirits, enthusiasm, dedication, determination, and laser-sharp focus.

The *really* good news is that you get to work *with* your SNPs, turning up the volume on your strengths and turning down the volume on your weak points. Through the Clean Genes Protocol, you can alter your lifestyle, diet, and environment to maximize the positives and mute the negatives, so that what you once thought was "normal" for you may be nothing of the kind. How awesome is that?

Some Genes Just "Act Dirty"

Sometimes a gene without a SNP creates problems for you anyway. That might be because your genes aren't getting the nutrients, lifestyle, or environment that they need to function at their best—too few vitamins, too little sleep, too many chemicals, too much stress. A better diet and lifestyle might inspire your genes to behave differently.

The scientific name for this is *genetic expression:* the way your genes express themselves in response to your environment, diet, lifestyle, and mindset. Depending on which of your genes are expressed, and how, you can be healthy, energized, and glowing. Alternatively, you might be loaded down with a whole slew of symptoms: obesity, anxiety, depression, acne, headaches, fatigue, achy joints, poor digestion. If your genes act dirty enough, you might even face such serious conditions as autoimmune disorders, diabetes, heart disease, and cancer.

Once again, your Clean Genes Protocol comes to the rescue. If you give your genes the diet and lifestyle that they need, they'll act clean instead of dirty, and you can optimize your health, your mental outlook, and your life.

Meet Your Dirty Genes

Here are the seven genes—I call them the Super Seven—that we target in this book. I chose them because they're extremely common, have been well researched, and have the most far-reaching effects on your body. If these guys are dirty—whether born dirty or just acting dirty—the rest of your genes will be gunked up, too. Some dirty genes are hard to scrub. Not these seven. They are easily cleaned up through diet and lifestyle changes.

Being born with dirty genes has an upside as well as a downside. Born-dirty genes might put you at risk for some nasty health challenges—but they also help to shape your personality, activating strengths as well as weaknesses. Your goal is to work with diet, chemical exposure, and lifestyle to maximize the benefits while minimizing the drawbacks.

① MTHFR, the methylation master gene

This gene initiates your ability to *methylate,* a key process that affects your stress response, inflammation, brain chemistry, energy production, immune response, detoxification, antioxidant production, cell repair, and genetic expression.

When MTHFR is born dirty:

Strengths: intensity, alertness, productivity, focus, improved DNA repair, decreased risk of colon cancer

Weaknesses: depression, anxiety, autoimmunity, migraines, increased risk of stomach cancer, autism, pregnancy complications, Down syndrome, birth defects, and cardiovascular conditions such as heart attack, stroke, and thrombosis

② COMT, the gene whose SNPs help determine whether you're focused and buoyant, or laid-back and calm

COMT and its SNPs have powerful effects on mood, focus, and how your body handles estrogen, a key factor in the menstrual cycle, in fibroids, and in some estrogen-sensitive cancers.

When COMT is born dirty:

Strengths: focus, tons of energy and alertness, good spirits, glowing skin
Weaknesses: irritability, insomnia, anxiety, fibroids, increased risk of estrogen-sensitive cancers, test anxiety, neurological disorders, migraines, PMS, impatience, vulnerability to addictions

③ DAO, the gene whose SNPs can make you supersensitive to certain foods and chemicals

When this gene is dirty, it affects your response to the histamine that lurks in various foods and beverages and that's also produced by some gut bacteria, affecting your likelihood of food sensitivities and allergic reactions.

When DAO is born dirty:

Strengths: immediate awareness of allergens and trigger foods (so you can get them out of your diet before they cause serious long-term problems)
Weaknesses: food sensitivities, pregnancy complications, leaky gut syndrome, allergic reactions, the risk of more serious conditions such as autoimmunity

④ MAOA, the gene that affects mood swings and carb cravings

This gene helps govern your levels of dopamine, norepinephrine, and serotonin: key brain chemicals that affect mood, alertness, energy, vulnerability to addictions, self-confidence, and sleep.

When MAOA is born dirty:

Strengths: energy, self-confidence, focus, "highs" of productivity and joy
Weaknesses: mood swings, carb cravings, irritability, headaches, insomnia, addictions

⑤ **GST/GPX, the gene(s) that can create detox dilemmas**
A dirty GST or GPX affects your body's ability to rid itself of chemicals.

When GST/GPX is born dirty:

Strengths: immediate awareness of potentially harmful chemicals (before they have the chance to make you really sick), improved response to chemotherapy
Weaknesses: supersensitivity to potentially harmful chemicals (with responses ranging from mild symptoms to serious autoimmune disorders and cancers), increased DNA damage (which increases the risk of cancer)

⑥ **NOS3, the gene that can create heart issues**
NOS3 affects your production of nitric oxide, which is a major factor in heart health, affecting such processes as blood flow and blood vessel formation.

When NOS3 is born dirty:

Strengths: decreased blood vessel formation (angiogenesis) during cancer, which reduces the growth of cancer
Weaknesses: headaches, high blood pressure, vulnerability to heart disease and heart attack, dementia

⑦ **PEMT, the gene that supports your cell membranes and liver**
This gene affects your body's ability to produce phosphatidylcholine, an essential compound that you need to maintain cell membranes, bile flow, muscle health, and brain development.

When PEMT is born dirty:

Strengths: more support for methylation, better response to chemotherapy

Weaknesses: gallbladder disorders, small intestine bacterial overgrowth (SIBO), pregnancy complications, cell membrane weakness, muscle pain

What Makes Your Genes Act Dirty?

Even if you don't have a SNP in any of your seven key genes, you might still be gunking up those genes with the wrong diet and lifestyle. As a result, they can't do the jobs you desperately need them to do—metabolize nutrients, balance your brain chemistry, repair damaged cells, and a hundred and one other tasks. What happens? You gain weight, feel sluggish, get depressed, become anxious, lose your ability to focus, develop acne, get headaches … the dirty list goes on and on.

If you're taking antacids, for example, you're messing with many major genes, including MTHFR, MAOA, and DAO. If you're taking metformin, a common medication for diabetes, you're disrupting the function of your MAOA and DAO. Birth-control pills, hormone replacement therapy, and even bioidentical hormones can strain your MTHFR and COMT.

Even if you're not taking medications, your genetic expression can be disrupted by poor diet, lack of exercise, too much exercise, not enough sleep, environmental toxins, and plain old everyday stress—and those are just the most common problems. Long story short, there's a whole laundry list of factors that might be dirtying up your genes—and your doctor probably has no idea.

To make matters worse, every additional factor that makes your genes dirty changes the whole picture. So if you're *only* eating too much sugar, that's one problem. But if you're *also* eating too many carbs, now you have two problems—and a much wider and more complex effect. If, in addition, you're not getting good sleep, you've just created more damage. That plus stress—even more! Pretty soon, you've generated a cumulative effect that makes the whole problem even worse. Instead of 1 + 1 + 1 + 1 = 4, you get 1 + 1 + 1 + 1 = 50.

Why? Because all your genes interact with one another. When one gene gets dirty, it doesn't work properly, so several more genes step up to help—and now suddenly they get dirty, too. Your body isn't a set of discrete compartments that each work separately. It's one amazing *interactive* system in which problems spread and multiply with amazing speed.

The good news is that health can also spread and multiply in amazing ways. When you clean up your dirty genes, you start feeling terrific in ways you never even imagined. Your mood improves—and that chronic muscle pain you've been working through stops aching. Your brain fog clears—and you've got tons more energy. Your allergy symptoms disappear—and you begin to lose some weight.

This is why I'm so eager for you to clean up your genes. If your dirty genes were born clean but became dirty, cleaning them up gives you a tremendous boost. And if some of your genes were born dirty, giving them the support they need can make a world of difference.

What Dirties Up Your Genes?

Diet

- Too many carbs
- Too much sugar
- Too much protein
- Not enough protein
- Not enough healthy fat
- A shortage of nutrients that your genes need to work properly, such as B vitamins, vitamin C, copper, and zinc

Exercise

- Sedentary lifestyle
- Overtraining
- Electrolyte deficiency
- Dehydration

Sleep
- Not enough deep, restorative sleep
- Going to bed late, getting up late
- Irregular sleep patterns

Environmental Toxins
- "Dirty" food
- "Dirty" water
- "Dirty" air—including indoor air
- "Dirty" products: sprays, cleaners, cosmetics, paints, pesticides, herbicides

Stress
- Physical stress: long-term illness, chronic infections, food intolerance/allergies, insufficient sleep
- Psychological stress: issues at work, at home, with your loved ones, with life

Our Toxic Environment

We're going to zero in on the role of chemicals in chapter 6. But they are such a big part of why your genes get dirty, I just have to say something about them now.

As an expert in environmental medicine, I am duty-bound to bring you the bad news: the industrial chemicals in our air, water, food, and products have gotten completely out of control. Our bodies were never designed to bear that chemical burden, and everybody's genes—no matter what they were born with—are staggering under the weight.

If only it were as simple as putting a filter on your water tap or sticking to the organic aisles in the grocery store! But what about the carcinogenic BPA coating that you rub onto your fingertips every time you accept a sales receipt? What about the formaldehyde fumes that rise out of your pressed-wood furniture, or the toxic perfluorinated compounds (PFCs) that live in your carpets? What about the chemicals exhaled by the photocopier at your office, or the bad effects of fluorescent lights and electromagnetic fields on your own biochemistry, or the thousand and one contacts you have with different types of plastic every day?

If it were just one, or two, or even twenty exposures daily, your body would have a much easier time shaking the toxicity off. But we're talking hundreds of daily exposures—maybe even thousands—that your body was never designed to withstand. Think of it: more than 120 million industrial chemicals are registered, and many of them end up in our air. With every breath you take, you're effectively bathing your body in a toxic soup. No wonder your genes are struggling!

I can tell you how to clean up the genes that have become dirty. And I can tell you how to give extra support to the ones that were born dirty. But sometimes I feel like one of those cartoon characters frantically trying to bail out a leaky boat that keeps springing leak after leak after leak. I keep trying to clean up those genes—and the chemicals in our food, air, water, and products keep dirtying them up again. Our factory-farmed food supply and chemical-laden environment are triggering the expression of many genetic problems that used to be silent and shutting down the expression of genes that used to function properly.

That's a huge part of why chronic disease is on the rise—such conditions as obesity, diabetes, heart disease, allergies, autoimmune disorders, and cancer. It's because the born-dirty genes that wouldn't have bothered you a hundred years ago, or fifty years ago, or maybe even twenty years ago are giving you trouble now. And even the genes that were born clean are becoming dirty at a disturbing rate. Even newborn babies are born with over two hundred chemicals in their bodies—on day one!

This is why I want you to eat organic or at least avoid the worst offenders. And to filter your water—what you drink, cook, and bathe in. This is why I don't want you putting chemical-laden products on your skin or washing your hair with them. It's why I want you to clean up the air inside your home, which is often more toxic than the air outdoors— strange, but true. And it's why I want you giving your genes every possible support you can: deep sleep, exercise that's right for your body, and stress reduction and relief.

You *can* take charge of your health. But if you feel like you're facing a massive challenge, that's because you are.

The Four-Week Clean Genes Protocol

Lucky for all of us, there *is* a way to support your genes, to scrub both those that were born dirty and those that are acting dirty. In just four weeks, you can go a very long way toward cleaning up your genes.

Step 1: Soak and Scrub for Two Weeks

In this phase, we clean all your genes.

- **Fill out Laundry List 1 (chapter 4): "Which of your genes need cleaning?"** You complete a questionnaire—a laundry list of symptoms and personality traits—that gives us your baseline. This helps us target which of your genes are functioning at less than their best, whether they were born dirty or are acting dirty.

- **Follow the program.** You follow a two-week program of healthy food, sufficient sleep, reduced toxic exposure, and stress relief. This part of the program is the same for everyone, because it's so good at clearing out the gunk. If your jeans were covered in mud and also had a few isolated grease stains, we'd have to soak and scrub all the mud away before we could zero in on the specific stains. This Soak and Scrub period of the Clean Genes Protocol works the same way. Supported by dozens of delicious recipes, you eat, sleep, exercise, detox, and destress for fourteen days.

Step 2: Spot Clean for Two Weeks

In this phase, we Spot Clean those of your genes that may have been born dirty.

- **Complete Laundry List 2 (chapter 14): "Which genes need more cleaning?"** You fill out a second laundry list to identify which genes are still dirty—perhaps because they were born that way, perhaps because they just need additional support.

- **Follow the program.** Now we make it personal. Based on your own laundry lists, you continue the Clean Genes Protocol for diet and lifestyle with specific adjustments for Spot Cleaning any genes that are still dirty.

Step 3: Keep Clean for Life

Throughout your life, you make sure that your genes stay clean and give any dirty ones the extra attention they need.

- **Complete Laundry List 2: "Which genes need more cleaning?"** Bring out that second laundry list every three to six months to target any dirty genes that are messing with you.

- **Follow the program.** Stick to the healthy diet and lifestyle you learned in your four-week program, bringing in your special Spot Cleaning techniques as you need them.

The Clean Genes Protocol: How You Can Clean Your Genes

Following is your Clean Genes Protocol—a lifelong program to keep your genes clean as you optimize your health. Although you may periodically add in Spot Cleaning from the second step of the program, this is the approach to diet and lifestyle that will best support your genes for the rest of your life.

In this chapter the Clean Genes Protocol is outlined in broad brush strokes, as an introduction. Fear not, though: in later chapters, we'll address all the components—diet, exercise, and so on—in greater detail.

Diet
- Eat appropriate amounts of protein and healthy fat.
- Make sure to get all the nutrients that your genes need to work properly, such as B vitamins, vitamin C, copper, and zinc.
- Cut out cow's milk dairy, gluten, excess carbs, and white sugar.
- Avoid foods high in pesticides, herbicides, preservatives, and/or artificial ingredients.
- Avoid fermented foods, leftovers, or food that's likely to contain excess bacteria if you find they trigger symptoms.

- Avoid foods that are high in histamines: wine, some types of cheese, and smoked and/or preserved meat and fish if you're particularly susceptible.
- Eat in moderation: stop eating when you're 80 percent full.
- Avoid snacks and late-night meals.

Exercise

- Get the right amount of exercise for your body—not too much and not too little.
- Exercise when you're rested and only until you're pleasantly tired. Don't exhaust yourself, and don't force it.
- Exercise when it doesn't negatively impact your sleep. Don't skimp on sleep to exercise; don't exercise later in the evening if it keeps you from falling asleep.

Sleep

- Make trying for deep, restful sleep a priority.
- Consistently match your sleep schedule to nature's circadian rhythms: asleep by 10:30 P.M., awake seven to eight hours later.
- Avoid electronic screens in the hour before bed.
- Block out or turn off artificial lights. Natural moonlight is great.

Environmental Toxins

- Eat organic food or at least avoid the "dirtiest" conventional foods.
- Filter the water you use for drinking, cooking, and bathing.
- Avoid the use of household and garden chemicals.
- Avoid all plastic containers for your food and water, especially BPA plastic and especially in the microwave. Ideally, store and cook foods only in glass or stainless steel.
- Follow guidelines to keep indoor air clean, bearing in mind that indoor air is often more toxic than the air outdoors.

Stress
- Attend to sources of physical stress: long-term illness, chronic infections, food intolerances/allergies, insufficient sleep.
- Reduce and relieve psychological stress: issues at work, at home, with your loved ones, with life.

If You've Had Genetic Testing . . .

What if you've had genetic testing already and want to head straight to "fixing" your problem genes.

Trust me: *Don't.*

If you're eating right, getting terrific sleep, avoiding toxins, and keeping stress at bay, your born-dirty genes—your SNPs—might not be giving you any trouble. How awesome is that?

And if you're *not* practicing an optimal diet and lifestyle, many of your genes are going to *act* as though they had SNPs even if they weren't actually born with them. What?! Yes! Being born without a particular SNP doesn't mean you're in the clear!

That's why your Clean Genes Protocol starts with the basics—a total Soak and Scrub for *all* your genes. Only then does it make sense to target any remaining problems as you maintain the diet and lifestyle that you learned during the Soak and Scrub. This Clean Genes Protocol has helped thousands of people worldwide. I want it to work for you, too. You'll get the best results if you follow the protocol, no matter what genetic testing you may have had.[*]

Genetic Testing: The Pros and Cons

Many of my clients have sent away for a genetic test from such companies as 23andMe and Genos Research. Sometimes the information is helpful, but often the results

---

[*] If your genetic testing reveals a serious medical condition—for example, cystic fibrosis, PKU, or the like—work with a physician to treat that condition and prevent or delay the worsening of symptoms.

can be confusing: "Take large quantities of vitamin X to support gene A; avoid vitamin X completely to support gene B; and consume moderate quantities of vitamin X to support gene C." How do you follow a recommendation like that? Unfortunately, most doctors aren't much help, either.

That's a big part of why I created this book—so you could clean up your genes without necessarily getting them tested. If you have a competent professional who can help you sort through the information, by all means, take the test. And if the Clean Genes Protocol doesn't give you the results you'd like, you might need to find a professional who can help you take things a step further.

In most cases, though, genetic testing isn't necessary. Just remain on the Clean Genes Protocol, and watch your health improve. Once you see how cleaning up your genes helps you eliminate fatigue, insomnia, irritability, ADHD, anxiety, depression, weight gain, and many other symptoms, you'll be ready to celebrate!

Why Do We Even Have SNPs?

From one point of view, SNPs are a real pain. Who would choose to have a gene that feeds your anxiety, encourages you to be obsessive, keeps you from falling asleep, or makes you supersensitive to toxins? Given a free choice, why wouldn't you always go for 100 percent clean genes?

But, as we have seen, dirty genes bring strengths as well as weaknesses. SNPs in the MTHFR gene, for example, can make you incredibly focused and determined to solve problems. COMT SNPs can give you tons of energy and buoyancy, enabling you to face life with an enthusiasm that many less energetic people might envy. GST/GPX and DAO SNPs alert you early on to the disastrous effects that certain chemicals and foods have on your body, leading you to make healthier long-term choices. Every dirty gene has its upside as well as its downside.

Scientists suspect that SNPs have an upside not just for individuals but for the community. Imagine a small band of early humans, trying to survive in the forest or on the tundra. Wouldn't it be useful to have one person who reacted especially intensely to potentially toxic foods, warning the rest of the group to stay away from those foods? Wouldn't you want one sort of obsessive guy who could never let anything go, so that the group didn't give up on solving problems, or one woman who was extra-alert to the sounds of danger just as everyone else was falling asleep?

Scientists also suspect that SNPs evolved because we humans live in so many different environments. Humans migrated all over the world, and in subtle but very important ways our bodies learned to adapt. SNPs may be part of that story.

Today, of course, you can obtain almost any type of food, no matter where you live, and if you need some extra support, you can supplement from the vitamin store. You don't have to be limited by your ancestors' environment—but you do need to understand how to respond to the SNPs you were born with. Fortunately, with the right information, you can support all your dirty genes, whatever they happen to be.

Success Stories: When Your Genes Get Clean

I encouraged Keri, Jamal, and Taylor to engage in the same four-week plan that you'll follow: a comprehensive program of diet, lifestyle, and prevention that can help you scrub your genes clean. All of them saw benefits—but at different speeds and in different ways.

For example, Keri responded very well to the Soak and Scrub phase of the protocol, but she was still bothered by a runny nose, scaly skin, and limp, lackluster hair. Clearly, her dirty GST/GPX gene needed more support. So we provided it:

- I advised her to take liposomal glutathione (an over-the-counter supplement) on a graduated dosage: she started with a tiny bit and moved up slowly in response to her symptoms. This supplemented her body's own inadequate production of glutathione, a key antioxidant.

- Although Keri had put filters on all her water taps, she had avoided buying an air purifier due to cost. Her continued symptoms made her realize that she was probably struggling with the dirty air in her home—fumes from furniture, carpets, cooking, her gas stove, her mattress, and other home products. To help offset some of these exposures, I advised her to use the vent hood while cooking and use high smoke-point oils. Dirty water and dirty air tend to be our two biggest exposures to problem chemicals, putting a huge burden on our GST/GPX genes.

- Keri began to take twice-weekly saunas so that she could "sweat it out." (I love saunas. Besides the fabulous stress relief, they're terrific for helping your body get rid of harmful chemicals.)

After just two weeks of Spot Cleaning, Keri's hair, skin, and energy levels showed huge improvement. In a few weeks more, her nasal passages had cleared up too, and she began to glow with health. Working *with* her genes instead of against them had made all the difference.

Jamal also benefited from the Soak and Scrub phase of the protocol, but he too needed some Spot Cleaning. He took liposomal glutathione, as Keri did, as well as a supplement called PQQ (more formally known as pyrroloquinoline quinone). I also advised him to increase his intake of arginine, a nutrient needed to support the NOS3 enzyme that the NOS3 gene produces. So Jamal began loading up his diet with arginine-rich foods: arugula, bacon, beets, bok choy, celery, Chinese cabbage, cucumber, fennel, leeks, mustard greens, parsley, and watercress. He began to lose weight, he felt better, and—perhaps most important—he was finally able to stop thinking that his genes had handed him a death sentence.

"Now that I know what to do, I feel like I can work *with* my genes, instead of worrying about what they're doing *to* me," he told me. "But I still need to help my father. I only wish I could have given my uncle and grandfather this information."

Taylor's depression and mood swings proved a bit more stubborn. Although she felt somewhat better after her two weeks of Soak and Scrub, she still felt flat and blah.

Taylor's MTHFR SNP could be supported by eating lots of leafy greens—salads, kale, collards, chard. But I knew that her depression would make it hard for her to get motivated. I didn't want to give her a task—like making salads or cooking greens—that would feel like one more burden. So instead, I started her on a twice-weekly supplement of methylfolate, one of the active forms of folate/vitamin B_9 (and a nutrient essential to the methylation process on which good health depends). I knew that once she felt better, she would have the energy to change her diet—and then we would likely decrease or even eliminate the supplement. I also had her take curcumin and PQQ, both of which would support her MTHFR gene and promote healthy methylation.

In addition, I urged Taylor to avoid packaged and processed foods, or at least to check the label for folic acid. Sometimes it seems as if folic acid is *everywhere*—it's a very common additive. Unfortunately, it blocks the pathways that methylfolate would normally use. With those pathways blocked, your body can't use methylfolate even if you're getting lots of it in your diet and supplements.

Sure enough, after a few days of taking methylfolate *and* avoiding folic acid, Taylor felt better; her depression lifted quickly. Over the next several weeks, she began making more salads and eating more greens, so we were able to adjust her dose of methylfolate down to once a week. As she continues to improve, we might be able to cut out supplements altogether.

"Wow," Taylor told me on her most recent visit. "What a difference! I feel like a whole new person." I could see it for myself: Taylor looked lively, enthusiastic, at peace with herself. By giving her genes what they needed, she had created a whole new life.

Reaching Your Genetic Potential

What Taylor, Jamal, and Keri achieved is what I want for you, too—using the Clean Genes Protocol to reach your genetic potential. I've seen this protocol work for scores of clients, and I've taught many physicians and health-care providers how to use it with their patients. It's the fastest, most efficient way to give your health a boost, and it's also a lifelong protocol that will keep your genes clean while optimizing your health. I'm thrilled to share it with you.

This field is expanding fast. I spend a lot of time keeping up with the latest research—but man, I absolutely love it! When a mother who has struggled with recurrent miscarriages comes up to me at a conference to show me her beautiful baby, or when a man writes me that for the first time in his life he feels free of depression and anxiety, I'm reminded all over again of how crucial it is to clean up our genes.

So let's get started! The next step is for you to learn some of the ABCs of genetic science—nothing heavy! Just some empowering knowledge that will help you on your way.

2

Gene Secrets:
What They Didn't Teach You
in Science Class

When Jessie and I began our session, she was brisk and impatient.

"My doctor tested me and found SNPs in my MTHFR gene," she told me. "What should I take?"

"Hold on," I answered her. "You don't have just one gene or one SNP—you have thousands of genes and possibly thousands of SNPs. And they're all talking to one another. We can't just zero in on one SNP. We have to look at the big picture."

Jessie looked confused. "I thought if I had a SNP, I was supposed to find the right supplement for it," she said.

I shook my head. "SNPs are important, and taking certain supplements to address certain SNPs might help. But one single SNP is rarely the *only* reason you're having health problems, so just taking a supplement isn't going to be enough. This approach is not 'a pill for an ill.' As I said, we have to look at things holistically."

Your Dynamic Genes: Part 2

In the Introduction, I explained that every moment of every day, your genes are giving instructions to your body. As you read these words, your genes are telling your metabolism to run fast or slow, which helps determine your energy level and your weight. They're giving instructions to your brain to regulate mood and mental focus, helping to determine whether you feel anxious or calm, depressed or optimistic, focused or scattered. Because your genes are always "talking" to your brain and body, you can't focus on only one small part of the conversation. You have to look at the *whole* conversation.

In this book, we're going to zero in on seven key genes that affect your brain and body. However, many more genes also play a role in your health, and *all* of your genes work together, in a dazzling array of interactions that happen so quickly and continuously, it's almost impossible to separate them. Okay—it *is* impossible to separate them. That's why any solution you come up with has to deal with the big picture.

Picture the rush-hour traffic in a busy city. Inching along among a crowd of belching cars, you're frustrated because you can't get from one end of town to the other—but what's your solution?

Suppose you form a citizen group and convince the city to block off one special avenue for crosstown traffic. Once drivers get onto that avenue, they've got an express lane all the way across town. Perfect! Well, no. While that sounded like a good idea before it was implemented—what about all the cars that ordinarily use that avenue for local errands? Now, with the new traffic pattern, they have to use other streets, and those other streets become even more crowded. You may have helped traffic on one street, only to create an even worse traffic jam everywhere else. For a solution to work, it has to address the *whole* problem, not just one piece of it, or you might make the overall problem worse.

It's the same with your body. There, the big picture involves adopting a diet and lifestyle that clean up *all* your dirty genes, and keep them clean.

That big picture also needs to factor in *haplotypes*—combinations of dirty genes. As you'll see in chapter 14, these combinations also play a powerful role in your health.

But first things first. One of our primary goals in the Clean Genes Protocol is to support methylation—one of the most important processes your body undergoes.

Methylation: Your Key to Clean Genes

Methylation controls your genetic expression. It determines whether a particular gene will be turned on—or off. *Every gene in every cell* is ultimately regulated by methylation.

Methylation involves adding a "methyl group"—one carbon atom plus three hydrogen atoms—to something (such as a gene, enzyme, hormone, neurotransmitter, vitamin) in your body.

When this occurs, we say that the chemical compound has been methylated.

What happens when this system fails? You have genes *on* when they're supposed to be *off,* and *off* when they're supposed to be *on.* A classic example is when methylation fails to turn off the genes that contribute to cancer. Not good.

Remember the "Tale of Two Mice" in the introduction? One mouse was overweight, bloated, and vulnerable to disease. The second was lean, energetic, and highly protected. As identical twins, the mice had the exact same genes. So what made the difference?

Methylation. The healthy mouse was *methylating* properly. The unhealthy mouse wasn't. If *you* are struggling with symptoms or disorders—anything from acne, headaches, and PMS to heart disease, diabetes, or obesity—you are almost certainly not methylating efficiently. And your health is suffering for it.

Love Your Liver

Some 85 percent of all methylation takes place in your liver, so you want to support your liver as much as you can:

- Drink alcohol in moderation—or, if you have methylation issues, avoid it altogether.
- Avoid industrial chemicals and heavy metals in your air, food, and water.
- Avoid unnecessary medications and recreational drugs.
- Help your body detoxify with the Clean Genes Protocol.

Methylation occurs countless times each second in every single cell of your body—so pause a moment, as you read this, and imagine all the methylation that's going on. Now consider the following processes—just a few of the hundreds that depend upon methylation:

Genetic Expression

Methylation turns off many of the genes that can otherwise lead to chronic conditions—the scary kind that run in families. Depression, anxiety, heart disease, dementia, obesity, autoimmune conditions, and cancer all have a genetic component. With proper methylation, you greatly reduce your chances of developing these conditions, because methylation literally alters the instructions that your genes send out. For example, genes that might loudly shout "Depression!" or "Heart disease!" suddenly become muffled or even silent when they're properly methylated.

Conversion of Food into Energy

If your body is good at converting food into energy, you eat less, maintain a healthy weight, and feel energized. If your body has trouble with this crucial process, you eat more, gain weight, and feel sluggish and exhausted. Many people struggle with the ups and downs of blood sugar. Their solution is to eat more often and to consume more carbohydrates. They soon become overweight and tired.

Methylation to the rescue—by helping to create a key compound called *carnitine*, which enables your body to burn fat as fuel. Now your blood sugar is more stable, plus you're *burning* fat instead of storing it.

Methylation also helps you burn fuel as efficiently as possible, with further benefits to your metabolism, your energy, and your weight.

Cellular Protection

Each of your cells is surrounded by a *membrane*, a wall that lets the nutrients *in* while keeping harmful elements *out*. To produce a strong cell wall, you need—you guessed it!—good methylation, which produces *phosphatidylcholine*, a key element in your cell walls.

Do you take vitamins? Supplements? Without proper methylation, they won't do you a bit of good. If your cell walls don't work right, the nutrients can't get into your cells; they just end up in your very expensive urine.

You also need phosphatidylcholine to regulate the rate of cell death and to make healthy *new* cells, to replace the 2.5 million that die every second. Without enough new cells, you might develop pain, fatigue, inflammation, and fatty liver.

Finally, you need phosphatidylcholine for your *bile,* a substance produced by your liver. Bile helps you absorb fat and regulates the bacteria in your small intestine. It flows from your liver into your gallbladder, so if your methylation is compromised, watch out for gallbladder problems.

Methylating for Two

During pregnancy, your body is methylating even more than usual to support the developing baby and its placenta. Nausea, vomiting, or gallbladder issues—all common during pregnancy—are frequently caused by poor methylation. And did you know that neural tube defects and congenital heart defects are not results of a folic acid deficiency, as we so often hear? They are results of a methylation deficiency.

Most health professionals aren't aware of this either—but now you are. So if you're pregnant or planning to become pregnant, make sure that you and your partner are both getting all the nutrients you need to methylate properly. The Clean Genes Protocol gives you a great start in that regard.

Brain and Muscle Health

Methylation also produces *creatine,* a compound that both your brain and your muscles use as fuel. If you've got muscular aches and pains, feel run-down and fatigued, or can't kick your brain into gear, poor methylation and low creatine might be the reason.

Production and Balancing of Neurotransmitters

Biochemicals such as serotonin, dopamine, norepinephrine, and melatonin are known as *neurotransmitters*—literally, the chemicals that help transmit messages throughout your body via *neurons.*

The right balance of neurotransmitters makes you feel sharp, clear-headed, focused, calm, optimistic, and enthusiastic. The wrong balance can make you feel foggy, confused, distracted, anxious, pessimistic, or simply blah. If you've ever struggled with anxiety, depression, brain fog, or ADHD, you know how important those brain chemicals are. And methylation is the key to great brain chemistry.

Stress and Relaxation Responses

Your *stress response* is produced by your *sympathetic nervous system,* which helps you to rise to the occasion: put in some extra effort, focus longer, work harder, and do whatever it takes to get a job done. Physical danger evokes the stress response—hence that response's nickname of "fight or flight." If your cave-dwelling ancestors saw a saber-toothed tiger, they had to either fight it or flee—and quickly! The need for demanding physical work likewise evokes the stress response—say, having to pull a fishing boat out of heavy surf, or facing a long trek across the desert in search of a new home. Other types of physical stressors that can set off a stress response include not enough sleep, an ongoing illness or infection, missing a meal, and taking a medication that's hard on your body.

Of course, emotional demands can also set off the stress response: a deadline at work, a crying child tugging at your sleeve, the prospect of dinner with a difficult friend or relative. Anything that creates a challenge for you—mentally, physically, or emotionally—triggers the stress response.

In all of these cases, your sympathetic nervous system produces a cascade of stress hormones—including epinephrine (adrenaline), norepinephrine (noradrenaline), dopamine, and cortisol—to help your body mobilize for extra effort. The stress response makes you feel alert, keyed up, ready for battle; perhaps you're breathing more quickly, and you have tenser muscles and a swiftly beating heart.

The stress response is ideally balanced by the *relaxation response,* produced by your *parasympathetic nervous system.* After any "fight or flight," it's time to "rest and digest." Your stress hormones subside, your tight muscles loosen up, your breathing becomes deeper, and your mental state shifts from "on alert" to relaxed and calm.

When you methylate efficiently, you've got the biochemicals you need to engage both responses. You gear up to meet your challenges during the day, and then relax to enjoy a peaceful evening before falling into a delicious sleep. You gear up to grind out a busy week, and then relax to enjoy a stress-free weekend. You gear up to make it through the busy season, and then relax on a two-week vacation.

When your methylation isn't working well, you don't always have the biochemicals you need for both sets of responses. You might feel permanently stressed, on a hair-trigger temper, unable to wind down, or so burned out that you can't find that extra oomph you need. Sure, your psychological outlook and your life circumstances are key factors. But a huge piece of the stress mess results from poor methylation.

Detoxification

Detoxification is your body's ability to eliminate chemicals that might be harmful, including industrial chemicals, heavy metals, and excess hormones. Your body needs a certain level of hormones, of course, but when those levels get too high, you can run into trouble.

Estrogen, for example, is an important hormone for both men and women. But when your body can't clear estrogen from your system, women run into issues with PMS, menstruation, menopause, and ovarian cancer, while both men and women are at risk for breast cancer.

To clear harmful chemicals and excess hormones, you need to methylate properly. Methylation also affects your ability to produce *glutathione*, your body's master antioxidant. It was lack of glutathione that caused Keri's symptoms in chapter 1, and support for methylation that helped her symptoms go away.

Immune Response

Your immune system is charged with attacking any "invader" that your body perceives as dangerous, including certain bacteria, viruses, and other *pathogens* (disease-causing microorganisms), as well as toxins, dangerous chemicals, and harmful foods. An underactive immune response makes you vulnerable to disease. An overactive immune response causes your immune system to attack not just hostile invaders, but your own tissue, eventually creating such autoimmune conditions as Hashimoto's thyroiditis (which attacks the thyroid), rheumatoid arthritis (attacks the joints), systemic lupus erythematosus (attacks the joints, skin, kidneys, blood cells, brain, heart, and lungs), multiple sclerosis (attacks the myelin sheath around your nerves), and many others.

Methylation helps your immune system find its sweet spot—neither too passive nor too active, but just right.

Cardiovascular Function

Faulty methylation can lead to atherosclerosis (hardening of the arteries) and hypertension, both dangerous to cardiovascular health. Excessive and/or chronic inflammation—which can be created by poor methylation—has also been implicated in cardiovascular disease.

DNA Repair

Your genetic instructions are embodied in your DNA, the deoxyribonucleic acid that is the biochemical code for life itself. Two strands of DNA, wrapped together in a spiraling double helix, contain the particular sequence of molecules that say who *you* are—and that tell your cells what to do to keep you alive and well.

Just as the couch in your family room is constantly undergoing wear and tear from being used so often, so is your DNA. Your body's own biochemical processes can damage your DNA, as can exposure to free radicals (unstable, highly reactive molecules), ultraviolet B rays (UVB), and certain biochemicals.

You want your DNA in peak condition so that it can give optimal instructions to every one of your cells. And guess what? Methylation is crucial to DNA repair, and it helps prevent DNA errors when new cells are built.

Marvelous Methylation

Methylation produces many key chemicals, including:

- Phosphatidylcholine
  - to produce cell membranes, enabling your cells to absorb nutrients and repel harmful ingredients
  - to produce bile, which helps you absorb fat and fat-soluble vitamins, and which keeps excess bacteria from growing in your small intestine
- Creatine, essential for brain and muscle function

- Norepinephrine and epinephrine, for energy, attention, and alertness
- Melatonin, for easing you into sleep
- Carnitine, for fat burning and energy
- Polyamines, for regulating your immune system

Methylation reduces many key chemicals, including:

- Histamine, which contributes to asthma, migraines and other headaches, insomnia, mania, allergies, and skin disorders
- Estrogen, which at high levels can contribute to acne, irritability, heavy menses, and cancer
- Dopamine and norepinephrine, which at high levels increase headaches, irritability, and stress
- Arsenic, high levels of which occur in many common foods and beverages (including water, apple juice, chicken, and rice), contributing to muscular weakness, tingling, and brown spots on skin

How Methylation Goes Wrong

I hope I've sold you on the importance of good methylation. So what interferes with good methylation?

No surprises here—our culprits are pretty much the usual suspects:

Poor Diet

When you're not eating the foods your body needs, your body can't methylate properly. First, you need protein, B vitamins, and a variety of other nutrients that methylation uses to produce the cells of your body and your brain.

However, the raw materials aren't enough. As you'll see in chapter 5, methylation is a complex sequence of biochemical events. Each of these reactions includes *cofactors:* vitamins or minerals that your body needs to spark the reactions.

Think of it this way: to get a bonfire going, you need some big logs, yes. But you also need kindling, tinder, paper, and a lot of matches; otherwise, those big logs are just going to sit there. The basic nutrients are the logs—but the cofactors are all the other key ingredients that enable the fire to start.

Folic Acid

Okay, just about everybody gets this one wrong, but I want *you* to get it right. The *natural* form of B_9 is called *folate,* and the active version of that natural form—that is, folate that's immediately usable by the body—is called *methylfolate,* a key compound for methylation. If you have any trouble methylating, you probably want to consume *lots* of folate, so that even if your methylation process isn't efficient, you'll end up with all the methylfolate you need. You find folate in leafy green vegetables such as spinach, mustard greens, collard greens, turnip greens, and romaine lettuce.

The *artificial* form of B_9 is called *folic acid.* It's found in vitamin pills and as an additive in many packaged foods. Folic acid is unnatural; it's not useful for your body at all until it's processed into an active, usable form.

However, because folic acid resembles folate, it gets into your *folate receptors,* where it *blocks* natural folate from getting where it needs to—inside your cell. As a result, if you're eating more foods containing folic acid than leafy green vegetables, the naturally occurring methylfolate struggles to get into your cells. And without enough methylfolate, your body can't methylate. In this way, folic acid *blocks* methylation.

It's important to understand this for two reasons. First, many doctors and other health practitioners *prescribe* folic acid, especially to pregnant women. *No. Stop.* If you need to take a supplement beyond what you can find in your diet, take *folate.* If it says "folic acid," put the package down.

Second, many types of conventionally produced foods are "enhanced" with folic acid. In 1998, in what I consider an almost criminally ignorant move, the Food and Drug Administration (FDA) began *requiring* U.S. manufacturers to "enrich" the following foods in that way:

- Bread   - Cornmeal   - Pasta   - Other grains
- Cereal   - Flour   - Rice

Now, if you're methylating well, and you're not eating huge amounts of the above foods, your body can compensate for a little bit of interruption by folic acid. But if you have dirty genes—either born dirty or acting dirty—you're probably *not* methylating well, and a big intake of folic acid is only going to make things worse. Folic acid is one of the worst methylation-blocking "vitamins" around.

The Wrong Amount of Exercise

We all know that exercise is good for us, right? One reason is because it supports methylation.

In a fascinating study conducted at the Karolinska Institute in Stockholm, scientists had young, healthy men and women bicycle at a moderate pace with only one leg while the other leg remained idle. After three months, they analyzed the DNA in each leg. The researchers found that more than five thousand sites on the muscle cells' genome showed new patterns of methylation—but only on the muscles of the leg that had exercised.

Exercise is better for methylation than no exercise—but *too much* exercise isn't good either. Why not? Because when you overexercise—exercise either too long or too intensely—you stress your body too much. And as you'll see below, excess stress disrupts healthy methylation.

Poor Sleep

When you don't get a good night's sleep, you don't methylate properly. And when you don't methylate properly, you don't make *melatonin,* a natural biochemical that helps you fall asleep—and stay asleep. It's a vicious cycle! Break it by getting some good sleep.

Too Much Stress

When your body is under stress, it uses up methyl groups much faster than when it's relaxed. To make more methyl groups, you need more methyl donors (found in the right foods and vitamins), and more energy. If the stress goes on long enough, you're likely to run out of methyl donors, energy, or both. You're no longer methylating properly, and your health begins to suffer.

Exposure to Harmful Chemicals

As Keri found out in chapter 1, chemical exposure can be overwhelming. If your body is staggering under a huge chemical burden, your genes are going to be frantically trying to compensate for the strain. Methylation will suffer as a result. And now, sad to say, you've got another vicious cycle:

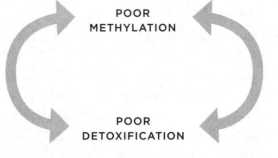

POOR METHYLATION

POOR DETOXIFICATION

Other Common Barriers to Methylation

- Alcohol
- Antacids
- Heavy metals
- Infections
- Inflammation
- Intestinal yeast overgrowth
- Nitrous oxide
- Oxidative stress (caused by free radicals)
- Small intestine bacterial overgrowth (SIBO) and other gut infections

---

Clean Genes and Good Methylation

Jessie, the client I introduced you to at the beginning of this chapter, was fascinated to discover the way her different genes all work together. She now understands that methylation is key to maintaining her genetic health (and thus her overall health), and that a variety of factors are needed to support methylation, including diet, exercise, sleep, protection from toxins, and stress relief.

I convinced her—and I hope I've convinced you. Your Clean Genes Protocol starts with a Soak and Scrub designed to clean up *all* your genes. Regardless of which genes you were born with, this overall Soak and Scrub is your best first step in supporting your health.

How Your Clean Genes Protocol Supports Methylation

- A diet rich in methyl donors and the vital nutrients that your genes need to complete methylation—without the packaged foods that contain folic acid
- The right type and amount of exercise
- Deep, restful sleep
- Avoidance of industrial chemicals and heavy metals; support for detoxification
- Stress reduction and stress relief

3

What's Your Genetic Profile?

Harriet was a bundle of energy—had been since she was a kid. Once she got started on a project, she hated to stop. Now that she was in law school, you could often find her up till the wee hours of the morning, wanting to read "just one more case" or finish "just one more page" before she finally felt ready to close her books.

Even after she had stopped studying, though, Harriet couldn't wind down. It might take her two, three, or even four hours from the time she stopped work to the time she fell asleep. "I can wind *up*, but I can't wind *down*," Harriet told me when we consulted about her sleep issues. "I've always been like that, but lately it's even worse. What's going on?"

Eduardo was the nicest guy in the world—but when he got angry, watch out! An intense man in his midforties, Eduardo had always taken life seriously, working hard to build up a small grocery business and proud of the way he had been able to take care of his aging parents, his three children, and his disabled sister. Eduardo believed deeply in family and was genuinely glad that he could offer support to so many people. He was disturbed, though, by the way his anger could, as he put it, "go from zero to sixty, without notice."

"I've always had a short temper," he told me, "but lately, it seems like the littlest things set me off. This morning, one of my boys spilled his juice—no big deal, he was already on the way to the sink for a sponge to mop it up. But before I knew it, I was yelling at the top of my voice how he should be more careful. I don't like being this way—but I can't seem to stop."

Larissa was the office manager of a small company, a job she had held for the past twenty years. In her midfifties, Larissa enjoyed her work, her family, and her hobbies, which included gardening and hiking with her family on the weekends. Larissa had always been calm and peaceful—one of those people others look to for a sense of perspective. "I have trouble getting too excited about anything," she told me. "It's just not my nature."

Lately, though, Larissa seemed almost *too* calm, to the point where she had trouble getting motivated or psyched about *anything*. "My husband suggested a family trip," she told me, "and I just couldn't find the energy to help him plan it. At work, it's the same thing—I just don't seem to care about solving the problems that are really my job to take on. It's as though everything has gotten a little dull, a little flat. Why is that?"

Harriet, Eduardo, and Larissa were all running into both the *upside* and the *downside* of their dirty genes. Harriet had been born with a slow COMT, which filled her with buoyant energy and good spirits—but also made it very difficult for her to wind down. When the rest of her genes were clean and Harriet was giving her dirty gene the support it needed, she could end her work at a reasonable time and get to bed at a reasonable hour.*

But now Harriet was under a lot of extra stress from law school. She wasn't eating a healthy diet or exercising properly, so her body was under both physical and psychological stress. As a result, *all* her genes had become dirty—and her dirty gene was acting even dirtier than usual.

Eduardo had been born with a dirty MTHFR. That genetic profile provided him with hard-driving determination and boatloads of motivation—but when his MTHFR got *too* dirty, Eduardo struggled with irritability and a hair-trigger temper.

---

* Sometimes a single gene can have two different types of SNP: one making it run too slowly, the other making it run too fast, each creating its own unique type of problem. COMT and MAOA are two genes of this type.

And, like Harriet, Eduardo was undergoing some extra stress lately. His daughter was having a rocky first year in high school, which was hard on the whole family. In addition, Eduardo had just battled the flu for a week, which put his body under considerable stress. Between his physical infection and his psychological stress, Eduardo's genes were getting dirtier by the moment—so his dirty MTHFR was giving him much more trouble than usual.

Larissa, in contrast to Harriet, had a *fast* COMT. In good times, this genetic profile gave her reservoirs of calm. But Larissa had been going through menopause, in her case a huge physical and emotional stressor. Under those conditions, her clean genes got dirty and her dirty gene became even dirtier. As a result, her inborn calm began to shade into a lack of motivation and drive.

Are you beginning to see a pattern?

When you support your genes with the right diet, exercise, sleep, protection from chemicals, and stress relief, your born-dirty genes are much more manageable.

When your body and/or mind undergoes stress, *all* your genes get dirty . . . and your born-dirty genes start to give you trouble.

The good news is that when you clean the dirt and gunk out of your system, you free up your genetic potential. That's why the Clean Genes Protocol works—it allows you to clean up *all* your dirty genes, and then Spot Clean the ones that need extra support. And if you maintain the Clean Genes Protocol as a lifelong approach to diet and lifestyle, you can keep your genes working at peak performance.

The Super Seven: A Snapshot

- MTHFR supports *methylation,* a crucial process that enables more than two hundred of your body's vital functions, including genetic expression.
- COMT affects metabolism of *dopamine, norepinephrine,* and *epinephrine,* affecting your mood, energy level, ability to calm down, ability to sleep, and ability to focus; it also affects *estrogen metabolism,* which governs your body's estrogen levels and hormonal

balance, affecting your experience of the menstrual cycle and menopause, and increasing your vulnerability to female cancers.

- DAO affects your body's response to *histamine from food and bacteria,* which in turn affects your vulnerability to allergy symptoms and food intolerance.
- MAOA affects your relationship to *dopamine, norepinephrine,* and *serotonin,* governing your mood, energy level, and ability to sleep, as well as sugar and carb cravings.
- GST/GPX enables *detoxification,* your body's ability to rid itself of harmful chemicals from the environment and to expel harmful biochemicals produced by your own body.
- NOS3 affects *circulation,* which helps determine your cardiovascular health and your vulnerability to heart attack, circulatory issues, and stroke.
- PEMT affects your *cell walls, brain,* and *liver,* determining a wide range of health issues including pregnancy problems, gallstones, fatty liver, digestive problems, SIBO, attention problems, and menopause.

What Dirty Genes Can Teach You About Yourself

For all three of the clients mentioned in this section, learning about their genetic profiles was exciting. As Harriet put it, "Suddenly, the way I am makes sense!"

Indeed, there were solid biochemical reasons why Harriet had trouble winding down, why Eduardo's temper tended to flare, and why Larissa lacked motivation. These traits were the downsides that surfaced when their dirty genes didn't get the support they needed.

For Harriet, a slow COMT meant that her body was slow to methylate both estrogen and dopamine. Estrogen is a female hormone that both men and women have. Like most biochemicals, we want the right amount—neither too little nor too much. Harriet's genetic difficulty with methylating this particular hormone meant that high levels of estrogen tended to hang around in her body. The upside: glowing skin, good sexual function (low estrogen frequently leads to vaginal dryness and atrophy or stiffness), and a smooth transition into menopause (which can be challenging precisely because of low estrogen levels). The downside: a tendency toward PMS and a vulnerability to estrogen-related cancers, including ovarian cancer and some types of breast cancer.

Harriet's slow COMT gene was also sluggish in methylating dopamine, the brain chemical associated with excitement, enthusiasm, and high-intensity energy. As a result, more dopamine remained in Harriet's system, keeping her energized and enthusiastic more intensely and for longer periods of time than would be typical of people with a clean COMT gene. When I tell you that dopamine is the biochemical involved in the thrill of a roller-coaster ride or the excitement of winning a big contest, you can imagine the extra boost of enthusiasm that excess dopamine added to Harriet's personality. And when I tell you that dopamine is also the chemical induced by a cocaine high, you can see why Harriet had a natural reserve of energy—but also had a hard time winding down.

Harriet had always worked in long bursts and then crashed afterward, feeling exhausted and burned out. I told her that to some extent, this was a natural rhythm that she should embrace. As long as she ensured that her hard work was balanced by relaxation, she could make the most of her natural gifts.

The danger, though, was that she might push herself too hard. In law school, Harriet had been working almost nonstop, sacrificing sleep and allowing herself very little downtime. As always, the key for Harriet was to support all her genes—both the ones that had been born dirty and the ones that had been born clean. They all needed the right diet, exercise, sleep, protection from toxins, and stress relief. Otherwise, Harriet's dirty COMT was likely to spin even further out of balance, and Harriet would become perpetually (rather than simply occasionally) exhausted and burned out. On the other hand, if we cleaned up all her genes, then her dirty COMT would be less burdened and could once again become an asset rather than a liability.

Eduardo, too, was pleased to learn that his focus, determination, and temper weren't "random," but rather part of his genetic inheritance. Eduardo's born-dirty MTHFR gene meant that he faced special challenges methylating. Since good methylation relies on folate, Eduardo would have to load up on high-folate foods, including leafy green vegetables, asparagus, broccoli, beans, peas, lentils, seeds, nuts, squash, and other vegetables. During stressful times—whether the stress was emotional or physical—Eduardo might even need to take some supplemental folate. That way, he'd be able to maintain his determination and focus without becoming irritable.

After Eduardo shared that both the flu and his daughter's hardships had created extra stress for his body and mind, I described for him how the stress challenged his fragile methylation, leaving him with less methylfolate than he needed. That lack of methylfolate meant that he could no longer quickly reduce his dopamine or norepinephrine levels. No wonder his temper was on a hair trigger and his anger felt hard to control.

Like Harriet, Eduardo was glad to understand what was going on. I told him he could find new ways to cope with stress—get enough sleep, go for a run, take a fifteen-minute "quiet time" to unwind. Supplemental methylfolate could also help him through the stressful times, as could being extra careful about his diet. After all, when he was stressed, his body needed all the methylation support it could get!

It wasn't only Eduardo's temper that was at risk. SNPs in the MTHFR gene make a person more vulnerable to headaches, autoimmune disorders, and certain cancers. These were high stakes for Eduardo—but now he had the tools to manage his genetic profile.

Larissa had the opposite profile from Harriet. Whereas Harriet's dirty COMT was *slow*, Larissa's dirty COMT was *fast*. Harriet's slow COMT kept dopamine and estrogen in her system longer than usual. Larissa's fast COMT got dopamine and estrogen *out* of her system faster than usual. As a result, Harriet's dopamine and estrogen levels were high, whereas Larissa's were low.

Larissa's genetic profile had given her a calm, peaceful temperament, which she normally enjoyed. When she felt *too* calm, however, it was a sign that some of her genes were acting dirty and that her born-dirty fast COMT wasn't getting the support it needed.

In her early fifties, Larissa was beginning menopause, a time of hormonal shifts that for many women is very stressful. The stress led Larissa's dirty fast COMT to expel estrogen and dopamine from her system even more quickly than usual. Because of her unusually low levels of estrogen, Larissa struggled with menopause symptoms—hot flashes, insomnia, reduced sexual function. Because of her extra-low levels of dopamine, she lacked motivation and energy.

Fortunately, the Clean Genes Protocol was able to help Larissa also. Once we cleaned up all her genes and supported her fast COMT, Larissa was able to overcome her symptoms and regain her enthusiasm for the things that mattered to her.

Your takeaway?

Clean all your genes, all the time. Make it a daily ritual!

Find out which born-dirty genes might need some extra support, and give that support to them.

Profiles and Personalities

Any one gene is just a single factor in your genetic profile—let alone in your entire personality. But to give you some idea of how your genetic profile might help shape your temperament, here are some quick personality sketches that I've observed in conjunction with the seven key genes when they're dirty:

MTHFR

Some days you're blue and depressed, while other days you're anxious. On good days, your focus is great and you get stuff done. On bad days, you have performance anxiety, a hair-trigger temper, and/or headaches— or maybe you just feel grumpy. After eating a salad you tend to feel great, but you've never paid attention to that because, after all, it's just a salad.

COMT (Slow)

Man, you're on fire! ADHD?—not in this house. You're cranking away on several projects and already eager for the next one or five. As you lie down to sleep, you're still cranking away. After tossing and turning, you finally doze off, dreaming of tomorrow's tasks. Tomorrow arrives. Coffee is needed. Once again, you're off and running. You put pressure on yourself, and if you're not accomplishing what you need to, anxiety sets in, so you focus harder to get everything done. And you *do* get it done. Your colleague makes fun of you for working overtime on a particular project, and you snap at her. As usual, you're quick to be irritated. In addition, sometimes you have an extreme sensitivity to pain and can be plagued with headaches.

COMT (Fast)

Look at that blinking light! Did you see that dog over there? Man, I wish I could read a book, but I just can't focus. You're always jumping from one task to another, and it's hard to get much done. Friends have suggested that you might have ADHD. You also love shopping and buying new things! The problem is, you feel great buying them, but the next day the "shopping high" wears off, and you find that you need to buy something else or you start feeling blue. It's getting expensive and time-consuming. Oh—and hugs! They're awesome! The more hugs you get, the better you feel.

DAO

You are so tired of not knowing what you can and can't eat. One meal you're good, and the next you feel awful: throbbing head, irritable mood, sweaty body, racing heart, itchy skin, bleeding nose. Perhaps you've even spent a ton of money on food allergy testing—and found nothing! So frustrating. You keep limiting your foods one by one in hopes of identifying the culprit, but it's a never-ending battle.

GST/GSX

Ever since you figured out that chemicals and smells make you feel sick, you've been on a mission to get rid of such stuff from your home. That neighbor of yours is using scented dryer sheets—again! Those give you a headache within seconds. Your friends wonder why you're such a clean freak. But you know that you're tuned in and sensitive to these things because you have to be.

MAOA (Fast)

Carbs. CARBS! Pleeease, get me some of those! Your grocery cart looks like you work for the grain and chocolate industry! You feel so great eating carb-laden foods. You know you shouldn't, but when you don't, you feel blue. The problem is, eating carbs picks you up only for a moment or two; then you crash. So what do you do then? You eat more carbs. You try diet after diet, but they just make you feel depressed. You're sick of gaining weight, but you feel stuck in that pattern. You don't want to be on antidepressants, but you feel like you can't keep going this way.

MAOA (Slow)

You're easily startled and quick to become anxious or irritated. You can become aggressive and later feel bad for overreacting. You just can't seem to help it. You always have to watch out for headaches, especially when you eat cheese or chocolate and drink wine. Falling asleep at night is always tough, but once you do, at least you sleep soundly through the night.

NOS3

You're freaking out. Your dad, uncle, grandma, and grandpa all had significant heart problems when they were around fifty years old, and now you're getting there. Your doctors check your heart and say it looks okay—but are they checking everything they need to, or are they missing something? Your hands and feet are constantly cold, but your doctors say that's nothing to worry about. You need answers, because this family history is weighing on you.

PEMT

Ever since you switched to a vegan or vegetarian diet, you've felt just a bit off. Your mind isn't as sharp, you're forgetting things, and you've got aches and pains all over your body. As an omnivore, you felt good overall, though you did have some aches and pains. Your liver felt heavy then, and it still does—just under your right-side ribcage. Fatty foods just don't sit right with you either. Now your doctor has said that you have gallstones and need to have your gallbladder taken out. No! There must be a way to save it.

What's Your Genetic Profile?

If you want to know your own genetic profile, there are a few ways to go about it.

The most expensive route is to get yourself tested by a company like 23andMe or Genos Research. At that point, you'll know exactly where all your SNPs are—but you won't necessarily know what those results mean.

Another route is to invest four weeks in this book's Clean Genes Protocol. Most people I know, including health professionals, get genetic testing results back and focus only on the genes. The problem is, that genetic report is a piece of paper showing your genetic *susceptibility*—not your genetic *destiny*! In other words, your genetic profile is not *you*.

Most of the folks who send away for genetic testing are unaware that a gene born clean can easily become dirty. When they read that their MTHFR is normal, they celebrate instead of realizing that—due to diet and lifestyle—it might in fact be superdirty.

Even if your MTHFR *was* born dirty, you don't want to make the common mistake of thinking that you can target it with a magical methylfolate supplement and all will be well. Many of the people who send away for their genetic profile end up following simplistic instructions, thereby creating significant side effects and making themselves worse off than they were before.

Here's the bottom line: the only way to truly help your dirty genes is by remaining on the Clean Genes Protocol, a lifelong approach to diet and lifestyle. That's how I do it. That's how my family does it. That's how the doctors I've trained around the world encourage their patients to do it. The result? Healthier, happier lives. I'll tell you what I tell all of them: there are no shortcuts. The tortoise always wins the race.

So here's what you need to do. Complete Laundry List 1 in the next chapter to find out which of your genes need more support. Spend two weeks on the general Soak and Scrub process. Then complete Laundry List 2 to further target specific genes with Spot Cleaning.

Because your genes can give you problems whether they're born dirty or simply acting dirty, I want you to know how to keep all seven of these important genes clean and healthy, giving them all the support they need. Out of the roughly twenty thousand genes in your body, these seven are critical to optimal health, every day of your life.

When I say you're going to scrub your dirty genes clean, I'm not saying that you're going to change your basic personality or remove all medical risk. I'm saying that you'll learn to work *with* your genes, giving them all the support they need. That way, you can celebrate your unique temperament—and safeguard your health.

Every vehicle drives differently. By understanding your genetics, you gain the opportunity and ability to make choices that will give you a smooth and enjoyable ride throughout your life.

PART II

MEET YOUR
DIRTY GENES

4

Laundry List 1: Which of Your Genes Need Cleaning?

This is an exciting moment for you. You're about to run through your first Laundry List of symptoms so that you can see which of your genes might be dirty.

Remember, you don't know which dirty genes were *born* dirty and which are simply *acting* dirty. Before we can begin blaming things on our genes, we need to see if our lifestyle, diet, nutrition, mindset, and environment (both indoor and outdoor) are influencing their function.

Let's get started.

Laundry List 1

No one else will see your answers here. Only you. Be absolutely honest. The goal of this exercise is to identify which genes are dirty so that you can make significant and strategic changes for the better.

If you're like me, your first impression might be, "Oh, I'm a mess!" But do as I have learned to do—reframe that negative thought into one that's both more positive and more accurate: "Wow! I have so much potential that I didn't know I could tap into!"

Check each box if the condition has occurred frequently within the last sixty days or is generally true:

MTHFR

☐ I suffer from headaches.

☐ I sweat easily and profusely when exercising.

☐ I take supplements with folic acid and/or eat foods enriched with folic acid.

☐ I struggle with depression.

☐ I have cold hands and feet.

DAO

☐ I tend to suffer from one or more of the following symptoms after eating leftovers, citrus, or fish: irritability, sweatiness, nosebleeds, runny nose, and/or headache.

☐ I'm sensitive to red wine or alcohol.

☐ I'm sensitive to many foods or suffer from leaky gut syndrome.

☐ I generally feel better two or three hours after a meal as compared to twenty minutes after eating.

☐ I felt better during pregnancy and could eat more varied foods.

COMT (slow)

☐ I get headaches.

☐ I find falling asleep a challenge.

☐ I easily become anxious or irritable.

☐ I suffer from PMS.

☐ I'm sensitive to pain.

COMT (fast)

☐ I struggle with attention and focus.

☐ I'm easily addicted to substances or activities: shopping, gaming, smoking, alcohol, social media.

☐ I'm prone to feelings of depression.

☐ I often lack motivation.

☐ I feel an initial happy rush after eating lots of carbs or starchy foods, but feelings of depression return quite quickly.

MAOA (slow)

☐ I'm easily stressed, panicked, or made anxious.

☐ I find it hard to calm down after becoming stressed or irritated.

☐ I enjoy cheese, wine, and/or chocolate but tend to feel irritable or "off" after I eat them.

☐ I'm plagued by migraines or headaches.

☐ I have difficulty falling asleep; but when I do, I tend to stay asleep.

MAOA (fast)

☐ I fall asleep quickly but wake up earlier than I'd prefer.

☐ I'm prone to depression and a lack of desire.

☐ I find that chocolate gives me a great mood lift.

☐ I tend toward smoking or alcohol addiction (or excessive use).

☐ I achieve a better mood after eating carbohydrates, but that improved mood doesn't help my focus or attention.

GST/GPX

☐ I breathe air and drink water. (Yes, you read that right! This gene is at least a little bit dirty in all of us these days.)

☐ I'm sensitive to chemicals.

☐ I developed gray hair early.

☐ I have a chronic condition such as asthma, inflammatory bowel disease, autoimmune disease, diabetes, eczema, psoriasis.

☐ I have a neurological disorder that results in symptoms such as tics, tremors, seizures, or problems with gait.

NOS3

☐ I have above-normal blood pressure (higher than 120/80).

☐ I have cold hands and feet.

☐ I tend to heal slowly after an injury or surgery.

☐ I'm a type 2 diabetic.

☐ I'm postmenopausal.

PEMT

☐ I tend to have generalized muscle pain.

☐ I've been diagnosed with fatty liver.

☐ I'm a vegetarian/vegan, or I don't eat much beef, organ meat, caviar, or eggs.

☐ I have gallstones or have had my gallbladder removed.

☐ I've been diagnosed with small intestine bacterial overgrowth (SIBO).

Scoring

Create a separate score for each gene, awarding one point per question:

- 0 points: Excellent! This gene is likely quite clean and functioning well!

- 1 point: Quite impressive! Your gene needs a bit of attention, but most likely because of problems in other genes rather than this particular one.

- 2 points: This gene seems to be a bit dirty. Fortunately, the Clean Genes Protocol will be a good first step in cleaning out the gunk. Cleaning the rest of your Super Seven will also help this gene function better.

- 3–5 points: This gene is definitely dirty. Two weeks on the Clean Genes Protocol will give you a great start. When you run through Laundry List 2, you can see whether this gene needs some extra attention.

My Score

MTHFR _____	MAOA (fast) _____
DAO _____	GST/GPX _____
COMT (slow) _____	NOS3 _____
COMT (fast) _____	PEMT _____
MAOA (slow) _____	

Getting to Know Your Genes

In the next seven chapters, you're going to learn about each of the Super Seven key genes.

Whatever your score on Laundry List 1, I urge you to read every chapter. Don't skip a chapter just because, in your case, a gene turned up clean.

Why? Because, as I mentioned earlier, your genes work together. The better you understand each of these seven genes, the better you'll be able to support your entire body and your overall genetic health.

Besides, a gene that's clean now might get dirty later. I want you up to speed on each of these genes so that you can recognize quickly when any one of them is having trouble. That way, you can be proactive about your health and stay ahead of the game.

One final reason—it's interesting! I know I'm a science geek, but I just can't help it: these are the building blocks of life itself. Your genes give instructions every single second that shape your health, your body, your personality. When you get right down to it, your genes are *you*.

So turn the page and start learning more about yourself.

5

MTHFR: Methylation Master

My friend Yasmin was in her midforties and had struggled with depression her entire life. Nothing clinical, nothing that kept her from holding down a demanding job as a biomedical technician or marrying a great guy and raising two wonderful kids. But whenever I met up with her, she always seemed a little down, a little flat.

"How are you?" I might ask.

"I'm okay."

"Well, how's your day going?"

"It's okay."

"And the kids?"

"They're okay."

You get the idea.

Yasmin wasn't a client of mine, but she was very interested in my work, especially when I told her about the research I was doing with MTHFR, a gene that has an enormous impact on our physical and mental well-being. When I told her that MTHFR SNPs are very common and that I and all three of my boys have them, she decided to have her doctor check out just that one gene. Sure enough, she had two SNPs.

Because her MTHFR gene was dirty, hundreds of functions in Yasmin's body weren't being properly methylated. As you saw in chapter 2, methylation is crucial to your body's health. And as you will see in this chapter, MTHFR is crucial to the *Methylation Cycle*—the process by which your body's genes, enzymes, and biochemicals receive the methyl groups that they need to function properly.

Because of its importance to the Methylation Cycle, a dirty MTHFR will soon cause you problems like Yasmin's. Your energy level droops. Your mental outlook suffers. Your metabolism gets out of whack. Your hormones go nuts. Your heart struggles.

So, what's the solution?

Well, the first step, as I told Yasmin, is to eat a lot of leafy greens. Your MTHFR's job is to methylate folate / vitamin B_9, turning it into *methylfolate,* the biochemical needed to kick off the Methylation Cycle. If your MTHFR is dirty, however, it can't methylate all the folate you need, so your Methylation Cycle can't function smoothly.

Luckily, you can help your dirty MTHFR by eating foods that already contain a lot of methylfolate, easing its burden and supporting your cycle. Leafy greens are full of methylfolate, so I told Yasmin to eat more salads and lightly cooked greens.

Ideally, diet gives you all the nutrients you need. But if your genes are dirty—and especially if they've been dirty a while—they might need another boost. Yasmin had been depressed for such a long time that her methylfolate levels were likely to be extremely low. For that reason, I also suggested that she take a methylfolate supplement to jump-start her recovery.

Methylfolate is a very potent supplement—you can't just start taking it in big quantities and expect everything to go smoothly. Some people can get away with that, but others will have very disturbing symptoms, anything from persistent anxiety to intense feelings of anger and aggression. As I do with my clients, I had Yasmin start slowly.

Yasmin started the dosage just a week before she and her family took a week's vacation to spend some time with her parents. Next thing I knew, her mother was on the phone to me, with her dad on the other line.

"What have you done to our daughter?" they wanted to know. "She's so happy! She seems to really be enjoying life. When you ask her how she is, she tells you about all the things that are going well, that she's excited about. *This* is the happy person we always hoped she'd be! What happened?"

When Yasmin came back, I could see the difference, too. She was still a quiet, thoughtful person, but there was an extra spark. Her affect was no longer flat; instead, it was lively and warm.

"I feel like I've come to life," she told me. "Can that one supplement really make such a difference?"

I told her I had seen the same type of response with dozens of patients, and that the doctors I'd worked with had shared hundreds of similar stories. In time, I told her, she might not even have to take a supplement—she might be able to get the same effect entirely from diet. That's the power of cleaning up your genes—especially your MTHFR.

Your MTHFR in Action

Dirty MTHFR genes are probably the most common of all the SNPs. You've already taken Laundry List 1, so you have a good idea whether your MTHFR is dirty, but here are a few more ways to track down a dirty MTHFR:

- I'm hypothyroid.
- My white blood count (WBC) has been on the low end of the normal range most my life.
- I get strong side effects from laughing gas (nitrous oxide).
- I had to have IVF or significant interventions in order to become pregnant and go to full term.
- One or more of my children are on the autism spectrum.
- One or more of my children have Down syndrome.
- My doctors say I don't tolerate medications such as methotrexate, 5-fluorouracil, or phenytoin as well as other patients.
- I get menstrual cramping and have clots in my menstrual fluid.

- My homocysteine levels are routinely high—above 12 micromoles per liter.

- My folate and/or vitamin B_{12} levels are elevated.

- I can't tolerate alcohol of any type well.

- I don't eat leafy green vegetables every day.

- I feel noticeably better after eating leafy greens.

MTHFR: The Basics

Primary function of the MTHFR gene

The MTHFR gene initiates the Methylation Cycle, a process that provides methyl to at least two hundred functions in your body.

Effects of a dirty MTHFR

Your entire Methylation Cycle is disrupted, affecting antioxidant production, brain chemistry, cell repair, detoxification, energy production, genetic expression, immune response, inflammation, and many other crucial processes.

Signs of a dirty MTHFR

Common signs include anxiety, brain fog, chemical sensitivity, depression, irritability, and a hair-trigger temper.

Potential strengths of a dirty MTHFR

Potential strengths include alertness, decreased risk of colon cancer, stellar focus, good DNA repair, and productivity.

Meet Your Dirty MTHFR

Whew—I know this one better than I'd like to! I speak from personal experience to say that those of us with dirty MTHFRs can be blue and depressed some days, while other days we're anxious. Yep, it alternates, and you never really know what's going to happen next, or why. MTHFR runs in families, so if you've got this particular dirty gene, your family members are probably also prone to mood swings. A dysfunctional MTHFR can lead to a whole slew of health problems.

Luckily, there's an upside. On good days, our focus is great and we get stuff done. We can drive through a ton of work, maintaining concentration from first to last. It's a blessing—but sometimes a curse as well, because we find it easier to rev up than to power down. And sometimes our families might wish we weren't quite so single-minded about *everything*, from completing a task to finishing an argument!

If your MTHFR was born dirty, it can have more than one hundred SNPs. However, lab companies test only the most common ones, so your test results are likely to show just one to four SNPs, which typically lower function to anywhere between 30 percent and 80 percent. (I'm at the low end, with only 30 percent function.)

However—and I want to be *super*clear about this—even if you were born with a 30 percent MTHFR, you might experience absolutely no symptoms at all.

Why not?

By now you can probably guess the answer:

When all your genes are as clean as they can be, your born-dirty genes will give you a lot less trouble—maybe even no trouble at all.

Don't believe me? Research proves it. Italians have a high rate of a type of SNP that reduces their MTHFR function to 30 percent. Most of them don't supplement with B vitamins even when pregnant. Yet Italians don't have children born with the birth defects typical of MTHFR SNPs.

Why? Because they eat their leafy green vegetables (diet), interact often and intimately with family and community (stress relief), and live in a generally beautiful, sunny climate (more stress relief). Their foods have not been factory-farmed and their dairy products are hormone-free (protection from toxic exposure). In other words, they live their lives following the principles that underlie the Clean Genes Protocol—principles that support healthy methylation. They're like the healthy mouse in "A Tale of Two Mice," using diet and lifestyle to erase any negative consequence of genetics.

Health Conditions Related to a Dirty MTHFR

Following are some of the disorders that researchers have associated with MTHFR SNPs. However, remember those Italians, who are typically healthy *in spite of* their SNPs! Genetics is not destiny, and the right diet and lifestyle go a long way toward keeping you fully healthy.

General Conditions

- Alzheimer's disease
- Asthma
- Atherosclerosis
- Autism
- Bipolar disorder
- Bladder cancer
- Blood clots
- Breast cancer
- Chemical sensitivity
- Chronic fatigue syndrome
- Down syndrome
- Epilepsy
- Esophageal squamous cell carcinoma
- Fibromyalgia
- Gastric cancer
- Glaucoma
- Heart murmurs
- High blood pressure
- Irritable bowel syndrome
- Leukemia
- Male infertility
- Methotrexate toxicity
- Migraines with aura
- Multiple sclerosis
- Myocardial infarction (heart attack)
- Nitrous oxide toxicity

- Parkinson's disease
- Pulmonary embolisms
- Schizophrenia
- Stroke
- Thyroid cancer
- Unexplained neurologic disease
- Vascular dementia

Pregnancy- and Birth-Related Complications
- Cervical dysplasia
- Miscarriages
- Placental abruption
- Postpartum depression
- Preeclampsia

Birth Defects
- Anencephaly
- Cleft palate
- Congenital heart defects
- Hypospadias
- Spina bifida
- Tongue-tie

The Methylation Cycle: Part 1

I call MTHFR the "methylation master" because it's the gene that kicks off your Methylation Cycle. As you saw in chapter 2, more than two hundred of your body's vital functions rely on methylation—functions such as skin repair, digestion, and detoxification. In other words, those functions need methyl groups to operate the way they're supposed to. And where do they get those methyl groups? From your Methylation Cycle. Since the Methylation Cycle is such an important factor in keeping your genes and your body in good shape, I want you to understand how it works.

Think of those two hundred functions, or processes, as two hundred gardens located throughout your body. Just as gardens need water, so do those processes need methyl groups. The Methylation Cycle is like an irrigation system that draws water from a clean, clear lake and distributes it to all the gardens. If something blocks or disrupts or dirties the irrigation system, some or all of those two hundred gardens won't get the water they need. Likewise, if something blocks or disrupts or dirties your Methylation Cycle, some or all of your body's processes either won't get the methyl groups they need or won't be able to use them properly.

In assessing the effectiveness of your Methylation Cycle, there are two issues you'll want to look at:

- Are methyl groups distributed to all the processes that need them?

- Can each process, once in possession of needed methyl groups, *use* those methyl groups effectively?

What might keep your Methylation Cycle from effectively using the methyl groups it produces? It's a long list! Chemicals, dirty genes, lack of vital nutrients, leaky gut, chronic infection, and stress can all either *block* the Methylation Cycle or significantly *slow it down*. The Clean Genes Protocol will help you clear up all such problems—which is why it's your best route to cleaning up your genes.

The Methylation Cycle: Part 2

So, you need your Methylation Cycle to work efficiently. How exactly does that happen?

As noted above, it's all about passing on methyl groups to functions that need them. Your MTHFR gets the process going by passing a methyl group to folate. That methylfolate then interacts with another biochemical—homocysteine, for example—passing the methyl group to it. Now homocysteine is methylated and becomes methionine. This process continues from one biochemical to another, in a kind of "bucket brigade" of genes and enzymes and biochemicals. Just picture a bucket full of methyl groups being passed on and on through the Methylation Cycle.

Finally, the bucket ends up in the hands of a biochemical called *S-adenosylmethionine*, or *SAMe* (pronounced "Sammy"). That's the player that ultimately passes those methyl groups on to the two hundred functions, or processes, that need them. When SAMe levels are too low or too high, those crucial processes in the rest of your body are greatly affected. Maintaining the right amount of SAMe is a fine balancing act, and MTHFR plays a crucial role in that process.

After SAMe has successfully passed on its methyl groups, it becomes a different biochemical, known as homocysteine. Homocysteine is the end product of methylation—but it's also the beginning: when your body is healthy and the Methylation Cycle is working well, homocysteine gets scooped up, methylated, and eventually turned into SAMe—and the whole cycle begins again.

Beautiful B_{12}

As we've seen, folate/B_9 and MTHFR are crucial to your Methylation Cycle. However, they can't do their jobs without the help of another B vitamin, called *methylcobalamin* (methylated vitamin B_{12}). Your Methylation Cycle depends on the teamwork of methylfolate and methylcobalamin. If either nutrient is deficient, your Methylation Cycle can't get a good start and those two hundred vital processes will never get the methyl groups they need.

The Methylation Cycle: Part 3

Once SAMe passes the methyl groups on to the various processes that need them, what happens next?

The binding of a methyl group changes many compounds in your body so that they have a new structure—and a new function. Sometimes the transformation occurs so that your body can *use* these new compounds. Sometimes it occurs so that your body can *expel* them.

Here are some examples:

Compounds Methylated to Use

- **Phosphatidylcholine.** Choline is a biochemical found in animal proteins. Methylate it, and you get phosphatidylcholine, which your body uses to make cell walls and to perform many other functions.

- **Creatine.** Methylated guanidoacetate becomes creatine, which is absolutely crucial for brain and muscle function.

- **Melatonin.** Methylated serotonin becomes melatonin, which you need to fall asleep.

Compounds Methylated to Expel

- **Arsenic.** When arsenic is methylated, it stops being active and your body can flush it out with the help of a superhero called glutathione.

- **Histamine.** Histamine is a powerful immune-system compound that you want in just the right amount. Too much gives you symptoms such as a runny nose or insomnia. When histamine is methylated, your body can expel it.

- **Estrogen.** Unmethylated estrogen is active, but methylated estrogen is expelled from your body. So methylation protects you from excess estrogen, which can cause PMS, menstrual issues, and the risk of estrogen-related cancers.

These are just a few of the critical biochemical reactions that depend upon SAMe.

Spectacular SAMe

You might have seen SAMe in the supplement store. Although it's sold over the counter in the United States, it's considered so potent in Europe that in Italy, Spain, and Germany you can get it only with a prescription.

And no wonder! Once you learn just how many chemical reactions depend on SAMe and the list of ills that SAMe is used to treat—stress, depression, anxiety, heart disease, gallstones, hepatitis, fatty liver, fibromyalgia, chronic pain, dementia/Alzheimer's, chronic fatigue syndrome, Parkinson's, multiple sclerosis, migraines, and PMS, to name only a few— you can understand how widespread its effects are.

Now, right away, a word of caution. Do *not* conclude that you need to rush out and take supplemental SAMe. Why not? My goal is not to load you up with extra compounds; on the contrary, my goal is to get your body to produce SAMe and support SAMe's role in the Methylation Cycle *on its own.* In addition, SAMe may not be right for you. (You will learn in later chapters whether you're a good candidate for supplemental SAMe.)

The Homocysteine Conversion

When SAMe has finished handing off its methyl groups, it turns into a brand-new chemical—one made only inside your body: *homocysteine.*

Think of a big sheet of cookie dough. You've cut out dozens of cookies in all sorts of shapes, but when you're done, some scraps of dough remain. That's homocysteine—the scraps left over after all that important methylating has gone on.

Your body has two choices for that leftover "dough": it can roll it back into the next batch of cookies, or it can use it for something completely different.

1. **The "more cookies" option.** Homocysteine is methylated and goes right back into the cycle.

2. **"Something completely different."** Homocysteine is used to make glutathione, a key detox biochemical. (Glutathione is so important that I often refer to it as a "superhero.")

The decision is easy for your genes. If your body is in good shape, not too stressed, everything working fine, homocysteine goes right back into the Methylation Cycle. More cookies!

But if you have a lot of free radicals and oxidative stress around— which happens if you've been skimping on sleep, feeling stressed, and exposing your body to a lot of toxins—then you're going to need more glutathione to clean things up. In that case, your homocysteine will be pulled out of methylation and put toward making glutathione.

This is why you want your body in good shape: you want to maximize the materials available to your Methylation Cycle, rather than shunting them off to make extra glutathione. The Clean Genes Protocol will help you do just that.

Measurement Myths: Your Homocysteine Labs

Because homocysteine is a product of the Methylation Cycle, many doctors believe that measuring your homocysteine levels is an accurate way of finding out whether you're properly methylating.

Sorry, it's just not that simple.

First, the levels of homocysteine that most doctors consider "normal" are actually way too high. Most doctors consider 15 micromoles per liter or higher to be high. For me, anything above 7 micromoles per liter is high. So if your doctor is measuring your homocysteine levels, make sure you get the actual number so that you can judge for yourself.

Second, sometimes your homocysteine levels are too low. If your homocysteine levels are *below* 7, you won't have enough homocysteine for both methylation and making glutathione. However, the labs don't necessarily tell you this—when you've got a low number, they just tell you that your levels aren't "too high," implying that they're fine!

Finally, you could have high homocysteine levels for lots of reasons—not only because you're methylating poorly. And you could have normal homocysteine levels and still be methylating poorly.

That said, you don't want your homocysteine levels to be above 7, because however they got that way, high homocysteine levels *block* your Methylation Cycle. The higher your homocysteine, the more your cycle is blocked. High homocysteine levels are associated with cardiovascular disease, neurological disorders, cancer, depression, anxiety, neural tube defects, congenital heart defects, cleft palate, infertility, and other issues—all of which result from a blocked Methylation Cycle.

Remember how I said that sometimes a person's Methylation Cycle can be blocked from *using* methyl groups even when they're available? This is a perfect example. If your doctor prescribes methylated supplements to reduce your homocysteine, yet your homocysteine levels don't go lower, that's a sign that your Methylation Cycle is blocked and is thus not able to utilize methyl groups as they're provided.

Such a blockage can occur for a number of reasons:

- Other dirty genes
- Inflammation
- Oxidative stress (free radicals)
- Heavy metals
- Folic acid (which blocks your folate receptors)
- Yeast overgrowth
- Small intestine bacterial overgrowth
- Infection
- Lack of needed nutrients

Fortunately, the Clean Genes Protocol will clear away these obstacles, ensuring both that you get all the methyl groups you need and that you can effectively use them.

Here's the bottom line: if your doctor wants to check your homocysteine levels to evaluate your cardiovascular risk, fine. But if his or her goal is to check your methylation, there are some tests that do a better job. (See Appendix A.)

What Makes MTHFR Dirty?

- Inadequate methylfolate (methylated vitamin B_9), methylcobalamin (methylated vitamin B_{12}), or riboflavin (vitamin B_2)
- Exposure to industrial chemicals
- Psychological stress
- Physical stress
- Hypothyroidism
- Folic acid

Riboflavin: A Crucial Nutrient

Riboflavin is crucial to the function of your MTHFR gene. Without it, your MTHFR can't function properly. What's more, a dirty MTHFR needs even more riboflavin than a clean one.

Long story short: make sure you're getting enough riboflavin in your diet, through such foods as spinach, almonds, and liver. Otherwise, your MTHFR won't be able to initiate the Methylation Cycle, and your whole body will suffer.

Key Nutrients for a Healthy MTHFR and Methylation Cycle

Here are some of the key nutrients that your MTHFR and your Methylation Cycle need to work properly:

Riboflavin/B_2: liver, lamb, mushrooms, spinach, almonds, wild salmon, eggs

Folate/B_9: green vegetables, beans, peas, lentils, squash

Cobalamin/B_{12}: red meat, salmon, clams, mussels, crab, eggs (vegans and vegetarians, you guys have to supplement)

Protein: animal sources including beef, lamb, fish, poultry, eggs, and dairy; vegan/vegetarian sources including beans, peas, lentils, broccoli, nuts, seeds

Magnesium: dark leafy greens, nuts, seeds, fish, beans, avocados, whole grains

---

Reasons You Might Be Low in Cobalamin/B_{12}

- Vegan/vegetarian diet
- Omnivore diet lacking enough meat, poultry, eggs, and fish
- High stress

- Antacid use
- *Helicobacter pylori (or H. pylori)*, a type of bacterium that can proliferate in the intestine
- Pernicious anemia (an autoimmune disease)

Methylation Miracles

I love reading about how life-changing clean genes can be. It makes everything I do seem worthwhile. I had inspiration for days after receiving the following letter from Sheryl Grelyak:

I never believed that anything could help my high-functioning autistic son. He had such severe anger and behaviors that looking forward I saw jail for him as it kept getting worse. As he aged into his late teens, he was suicidal and depressed, and by nighttime it was horrific—10 P.M. was the witching hour.

When his psychiatrist happened to do a gene test, I stumbled upon the fact that my son had MTHFR SNPs. I started your protocol, which I read about on the Internet, and I cannot tell you what a radical change it made in my son—immediately. He has bathed every day for sixty days, he is less angry, and there are no more issues with depression or suicide. When I say, "Time to get off the Xbox," he's like, "Okay, Mom." He's happy, and I now see independence in his future. He no longer gets so angry that he's violent or scary. It's like a miracle.

I will tell you that I knew nothing, absolutely nothing, about genes—I'd never even heard about SNPs until September—so I'm not one of the fanatical people. I'm a regular mom who stumbled on genetic help for my son's problems due to one test by a doctor who knows nothing about MTHFR. All I did was follow your protocol from your website. Before we started your protocol, my son had been through every school in our district. He's at a nonpublic school as a last resort, and the district told them to start calling the police if there were further problems. I mean, we were at our breaking point. Then you came along, and wow wow wow. It's been almost seven months and he's doing so well! It's been a miraculous change.

My daughter, who is in nursing school, has a dirty COMT and MAOA. Once we found out, we started her on SAMe, and it made a huge difference—she's bubbly, happy, and truly a new gal. She is amazed.

You have made a great impact on my children's lives without ever meeting them. Thank you.

Methylfolate: A Potent Supplement

I've talked about methylfolate on earlier pages—how important that nutrient is and which foods are rich in it. Some people, assuming that more is better, want to jump right to *supplemental* methylfolate.

"A pill for an ill" is a powerful notion—but as you know by now, it's not how I roll. You *might* need to take supplemental methylfolate—or you might be able to get the same results through diet and lifestyle alone. Even if you do need to take methylfolate (that is, if you try diet and lifestyle and can't get your MTHFR clean), the supplement might have no effect—or even a negative effect—depending on a host of factors.

We'll get into all that in the Spot Cleaning chapter, and I'll talk you through everything you need to know. Till then, please keep reading, do your two weeks of initial Soak and Scrub, and wait to find out what's right for *your* body.

The Choline Shortcut

If your body doesn't have enough methylfolate (methylated B_9) or methylcobalamin (methylated B_{12}) to manage the full Methylation Cycle, it recognizes the problem and takes what I like to call the *Choline Shortcut*.

This shortcut is very common in people with a dirty MTHFR, since most are short on methylfolate.

This shortcut, like the Methylation Cycle, methylates homocysteine, but instead of relying on the B vitamins for that task, it relies on the nutrient *choline*, which is found in eggs, red meat, poultry, fish, caviar, liver, and other organ meats.

Although you *can* get choline from some vegetables, such as spinach and beets, most vegans and vegetarians are deficient in it, just as they tend to be deficient in B_{12}. Vegans and vegetarians, you guys *do* need to supplement with choline and B_{12}, regardless of whether you have a dirty MTHFR or not.

Now, the Choline Shortcut may work for a short time, but you can't rely on it permanently. It's an emergency shortcut that your body follows to protect your liver and your kidneys. Your primary Methylation Cycle, by contrast, supports *all* your organs and tissues, including your brain, eyes, uterus (and placenta), testicles, skin, and intestines, to name only a few. The shortcut route doesn't suffice for them.

Getting enough choline from your diet is important, but first and foremost, support your Methylation Cycle.

Making the Most of MTHFR

As I lay out the Clean Genes Protocol, you're going to learn everything you need to know about how to support your MTHFR gene, whether it was born dirty or is only acting dirty. Luckily, I've had years of personal practice in supporting this gene, for me and for my three sons (as well, of course, as for my many clients). I'll happily share with you our keys to balance.

Meanwhile, if you think your MTHFR was born dirty, here are a few suggestions to get you started. You can get moving on the following suggestions as soon as you like, even before you start the Soak and Scrub:

- Know that your moods will naturally ebb and flow, and don't let mood swings throw you. Recognizing your varied nature will help you be more okay with feeling blue one day and anxious the next.

Our goal is to get you more of those good focused and productive days—and we can.

- Folic acid is your enemy. And it's everywhere—supplements, energy bars, foods, drinks. Cut it out of your life. Right now.

- Filter your drinking water. By removing arsenic, chlorine, and other unwanted chemicals from your water, you reduce the work your dirty MTHFR has to do.

- Leafy green vegetables are key for you. Eat them. Often.

- Make sure you're getting enough B_{12}. Eat sufficient beef (grass-fed only), lamb, eggs, crab, clams, and dark-meat fish. Vegetarians and vegans need to check out the Clean Genes Protocol for instructions on how to ensure adequate dietary choline and/or methylated B_{12}.

- In many cases, you'll want to avoid cow's milk dairy entirely. Food allergies and/or sensitivities to dairy produce antibodies that can clog your folate receptors. Goat's and sheep's milk products are usually fine— unless you have an autoimmune disease—and cow's milk products *may* be okay after you've cleaned up your diet and healed your gut.

The Trouble with Serum Folate Tests

Your doctor might order a serum folate test that supposedly measures your folate levels. But "folate" is a slippery term, as we saw earlier. In a serum folate test, a lab is actually measuring *both* artificial folic acid (from supplements and "enriched" foods) *and* natural folates (from whole foods). The lab results don't tell you which is which.

The moral of the story? *Ignore* any standard lab reading of "serum folate" if you're taking any type of folic acid supplement—either separately or in your multivitamin—or if you're consuming significant amounts of folic acid in your food. (See chapter 2 for a list of foods "enriched" with folic acid.) These lab readings are meaningful only if you're *not* consuming folic acid. (I provide additional information about folate testing in Appendix A.)

How the Clean Genes Protocol Supports
Your MTHFR and Your Methylation Cycle

Diet. By loading up on leafy greens, you compensate for the methylfolate that a dirty MTHFR can't easily produce on its own. By avoiding folic acid, you ensure that your folate receptors remain open to methylfolate and keep your MTHFR as clean as possible. By getting a range of other nutrients—especially B vitamins, protein, and magnesium— you're ensuring that your Methylation Cycle has all the nutrients it needs, which is crucial both for your MTHFR's function and for your overall health.

Chemicals. By avoiding exposure to industrial chemicals and heavy metals, you're keeping your MTHFR as clean as possible while significantly easing the burden on your Methylation Cycle; in addition, to the benefit of your overall health, without those chemicals your body can methylate more homocysteine, because less is needed for glutathione. By reducing or avoiding alcohol and avoiding nitrous oxide, you're also keeping your MTHFR as clean as possible while easing the burden on your Methylation Cycle.

Stress. Deep, restorative sleep is your best friend here— it's the best stress-reducer I know! Along with other types of stress relief on the Clean Genes Protocol, sleep will lighten the burden on your MTHFR as well as on your Methylation Cycle.

6

COMT: Focus and Buoyancy, or Mellowness and Calm

When Margo and I met, her exuberant personality seemed to fill the room. She smiled at me enthusiastically, but she looked tired and drawn, though she was only in her midthirties. And when she began reeling off her long list of symptoms, I understood why. She had trouble sleeping. Caffeine made her more irritable and anxious, though those feelings plagued her even without it. She got fierce headaches every month, the day before her period. She had a demanding job as a high school administrator, which she loved, but which left her feeling burned out and exhausted at the end of every week. As she put it, "If it weren't for weekends to recover, I don't think I'd be able to start up another week. I love my weekends."

Margo's personality and health assessment fit the profile of "slow" COMT SNPs so perfectly that I wasn't surprised when her test results came back. I explained to her that her particular inheritance of a dirty gene was loaded with both pros and cons:

Strengths

- Natural enthusiasm and exuberance

- Altruism and generosity

- Energy and productivity
- The ability to focus for long periods of time

Weaknesses

- Trouble winding down
- Sleep challenges
- Workaholism
- Difficulties metabolizing estrogen (which can lead to menstrual issues, fibroids, and female cancers)

When I met Blake the following hour, I was struck by the ways in which he was almost a mirror image of Margo. A carefree young man in his late twenties, Blake was the ultimate laid-back personality. He slept deeply—but even after a good night's sleep, he rarely felt energetic. He loved his coffee—a few times a day to energize him. He had lots of interests—world music, Japanese literature, exotic reptiles, horseback riding—but he had trouble focusing on any one activity for very long. Although he was deeply committed to his friends and his girlfriend, he told me that it was hard for him to show up for their dates on time and that sometimes he even forgot plans completely. "Something else comes up," he explained with a shrug, "and then I just get involved in that."

If Margo fit the profile of the slow COMT, Blake was the poster child for the fast COMT. Like Margo, Blake's dirty gene provided him with both strengths and weaknesses:

Strengths

- Natural calm and ability to relax; high tolerance for stress
- An undemanding and accepting nature
- A broad focus; a wide range of interests
- An ability to sleep well

Weaknesses

- Trouble revving up

- Difficulty maintaining focus; distractibility
- Weak memory
- Tendency toward depression

Both Margo and Blake had dirty COMT genes. But while Margo's dirty gene was too slow, Blake's was too fast. Each responded in different ways to their body's biochemicals.

Your COMT in Action

The COMT gene determines your ability to process catechols, estrogen, and some major neurotransmitters—dopamine, norepinephrine (noradrenaline), and epinephrine (adrenaline). *Catechols* are compounds found in green and black tea, coffee, chocolate, and a few green spices, such as peppermint, parsley, and thyme; as well as in the supplements EGCG, green coffee-bean extract, and quercetin. *Neurotransmitters* are the biochemicals that, in the brain, enable us to process thought and emotion. Let's look at three key neurotransmitters:

Dopamine

Dopamine is a neurotransmitter involved in excitement, thrills, and uncertainty. A burst of dopamine is a huge reward; it makes you feel terrific! When I tell you that falling in love is accompanied by a huge dopamine rush, you can see how good it makes you feel. Dopamine is also present in high levels when you're gambling, riding a roller coaster, or getting ready to meet a big challenge—any high-stakes activity where the outcome feels uncertain.

Likewise, dopamine is also involved in addictions. The dopamine rush following the ingestion of certain drugs is so pleasurable that you feel as though you'd do anything to repeat it. Cocaine, for example, triggers dopamine. But so does a thrilling first date, a scary movie, or skydiving. The technical name for this triggering of dopamine and our urge to repeat it is "reward system." Dopamine is your body's ultimate reward—the thing that feels so good, you'd do most anything to feel it again.

Norepinephrine and Epinephrine

Norepinephrine and epinephrine are your two key stress neurotransmitters. They help you rev yourself up for big challenges—anything that requires extra physical or emotional effort. If you're an emergency room doctor, nurse, or orderly, for example, you might need repeated jolts of norepinephrine and epinephrine all through your shift to keep jumping into action each time a new patient is wheeled through the door.

As you saw in chapter 2, your body has both a *stress response* to meet new challenges and a *relaxation response* to rest, heal, and recover. Ideally, both are balanced, so you rev up quickly for a big challenge, meet the challenge, and then restore your energy with a calm meal and a good night's sleep. How easily you rev up depends at least partly on how quickly your body can pump norepinephrine and epinephrine *into* your system. How effectively you power down depends at least partly on how quickly you can clear those biochemicals *out* of your system so that you can relax and regenerate.

Margo's slow COMT was slow to clear catechols, estrogen, dopamine, norepinephrine, and epinephrine from her system. As a result, her levels of these compounds tended to be high. The extra estrogen gave Margo glowing skin and good sexual function, but it also caused monster PMS and put her at risk for breast and ovarian cancer. The extra neurotransmitters gave her abundant energy, enthusiasm, and drive, filling her with a sense of confidence and optimism. But they also made it hard for her to power down, take a break, and get restful sleep. They also made it very difficult for her to calm down after caffeine—her dirty COMT had trouble clearing that stimulant from her system.

Only half-joking, I told Margo, "Most of the time, you're Superwoman. You've got tons of energy, drive, and focus—"

Laughing, Margo finished the sentence for me, "—but that one time of the month, *watch out.*" Indeed, Margo often found herself irritable and snappish, the result of all the extra estrogen and neurotransmitters.

Blake's dirty COMT—the mirror image of Margo's, as I noted earlier—was fast. It metabolized catechols, estrogen, and stress neurotransmitters so quickly that Blake's levels were usually low. Women with this gene struggle with low estrogen, which leads to such symptoms as vaginal dryness, reduced sexual function, and, in their fifties, trouble with menopause, when estrogen levels are dropping anyway. They're also at risk of heart disease.

Meanwhile, Blake's low levels of stress neurotransmitters gave him an enviable serenity and calm—an admirable ability to shake off the little irritants of life that so often bother the rest of us. Most things genuinely didn't bother Blake; he was wired for acceptance, adjustment, and compromise.

The downside was that he often lacked the ability to focus, bear down, and get things done. He didn't mind if you were an hour late to an appointment—but he didn't necessarily mind if *he* was, either. And since his dopamine levels tended to be low, he often lacked energy and confidence. "I do my best," he told me, "but I don't expect that much to come of it. Caffeine and chocolate are my go-to pick-me-ups. They help—but the boost doesn't last."

As you can see, both types of dirty COMT have their strengths and weaknesses—and both pose specific challenges to your health. Our goal, as always, is to support the strengths while minimizing the weaknesses.

COMT: The Basics

Primary function of the COMT gene

The COMT gene affects the way you metabolize estrogen, catechols from food and drink, and the stress neurotransmitters dopamine, norepinephrine, and epinephrine.

Effects of a dirty COMT

Slow COMT. You may not be able to clear catechols, estrogen, dopamine, norepinephrine, and epinephrine from your system. As a result, they remain in your system longer than they should, with a variety of physical and psychological effects.

Fast COMT. You clear catechols, estrogen, dopamine, norepinephrine, and epinephrine from your system *too* efficiently. As a result, they leave your system sooner than they should, with a variety of physical and psychological effects.

Signs of a dirty COMT

Slow COMT. Common signs include buoyancy, confidence, energy, enthusiasm, strong sexual function, estrogen issues (PMS, menstrual issues, fibroids, risk of female cancers), irritability, pain intolerance, sleep difficulties, trouble relaxing or powering down, workaholism, and sensitivity to caffeine, chocolate, and green tea.

Fast COMT. Common signs include excessive sense of calm, good-temperedness, lack of sleep difficulties, effective stress response, pain tolerance, difficulty completing tasks, difficulty focusing, forgetfulness, lack of confidence or optimism, low energy, menopause/perimenopause challenges, and reliance on caffeine, chocolate, and green tea.

Potential strengths of a dirty COMT

Slow COMT. Potential strengths include altruism, energy, enthusiasm, exuberance, focus, generosity, and productivity.

Fast COMT. Potential strengths include ability to relax, acceptance of others, broad focus, calm, high tolerance for stress, restful sleep, and a wide range of interests.

Meet Your Dirty COMT

You've already run through Laundry List 1, so by now you have a sense of whether you have a dirty COMT. But just to help you paint a fuller self-portrait, here are a few more traits associated with slow and fast COMTs. Do you recognize yourself in either category?

Slow COMT

☐ I've always been able to focus and study for long hours.
☐ I enjoy traveling and exploring.
☐ I tend to be a workaholic.
☐ When I get stressed, it takes me a long time to calm down.

☐ I tend to work hard for weeks, then crash and need to take a long break to recharge.

☐ I get anxious and panicky easily.

☐ I find that caffeine often increases my stress.

☐ I'm easily irritated. I wake up on the wrong side of the bed often.

☐ I have strong bones.

☐ It takes me a long time to fall asleep.

☐ My skin glows and people compliment me on it.

☐ I had early menarche.

☐ I usually get PMS.

☐ I experience heavy menstrual bleeding (menorrhagia).

☐ I have or have had uterine fibroids.

☐ I'm sensitive to pain compared to others.

☐ Eating a higher-protein diet (such as Gut and Bowel Syndrome [GAPS] or Paleo) makes me feel more irritable.

☐ I do poorly with central nervous system (CNS) stimulant medications such as Ritalin, Adderall, Vyvanse, and Focalin.

☐ I do better with CNS calming medications such as Intuniv.

Fast COMT

☐ I have a difficult time paying attention. I'm a poster child for ADHD.

☐ I tend to go with the flow.

☐ I'm not a workaholic.

☐ When I get stressed, I recover quite quickly and move on.

☐ I fall asleep quickly.

☐ Where's my cup of coffee? I need it!

☐ Eating a higher-protein diet (such as GAPS or Paleo) makes me feel great.

☐ I tend to be more depressed than enthusiastic, and have been for years.

☐ I'm just not that excited about things.

☐ I had late menarche.

☐ I don't get PMS.

☐ I have (or had) typically light menses.

☐ I have weaker bones.

☐ I'm very tolerant of pain compared to others.

☐ I do better with CNS stimulant medications such as Ritalin, Adderall, Vyvanse, and Focalin.

☐ I do worse with CNS calming medications such as Intuniv.

Health Conditions Related to a Dirty COMT

Whether your COMT was born dirty or is just acting dirty, it can create problems for you if you don't clean it up. Following are some of the disorders that researchers have associated with a dirty COMT.

Slow COMT

- Acute coronary syndrome
- ADD with hyperactivity
- Anxiety
- Bipolar disorder—especially mania
- Breast cancer
- Fibroids
- Fibromyalgia
- Panic disorder (especially in women)
- Parkinson's disease
- PMS
- Preeclampsia
- Schizophrenia
- Stress cardiomyopathy
- Stress-related hypertension
- Uterine cancer

Fast COMT

- ADHD—inattention, multitasking, inability to focus
- Addictive disorders—whether to drugs, alcohol, gambling, shopping, or video games
- Depression
- Learning disability

The Methylation Cycle and COMT

By now, you've got a pretty good idea of how important the Methylation Cycle is. As you saw in chapter 5, the SAMe portion of that cycle hands off methyl groups to facilitate some two hundred different processes.

One of those methyl groups goes to an enzyme produced by your COMT gene, the COMT enzyme. When that happens, two processes should occur:

1. Estrogen is methylated and expelled from your body.

You need *some* estrogen, of course. Although men and women need different levels, you both need some. If you clear it *too* quickly, that over-efficiency causes your estrogen levels to drop too low. That's what may happen with a fast COMT.

On the other hand, you don't want it to linger and cause your estrogen levels to remain too high. That's what happens with a slow COMT. Your goal is to find the "Goldilocks speed" for estrogen elimination: not too fast, not too slow, but just right.

2. Your stress neurotransmitters are methylated.

   —Methylated dopamine becomes norepinephrine.
   —Methylated norepinephrine becomes epinephrine.
   —Methylated epinephrine is ready to be expelled from your body
     by another set of enzymes.

Dopamine, norepinephrine, and epinephrine are all stress neurotransmitters, as we've seen. They're designed to make you alert, focused, and ready to spring into action. The stress response gets you breathing more quickly, causes your muscles to tense up, and sharpens your mind. It also makes it harder for you to digest your food, have sex, conceive a child, or fall asleep.

The relaxation response, on the other hand, is supposed to counter those effects. Your breathing returns to normal, your muscles loosen, your mind relaxes, and you become ready to digest, have sex, conceive, and sleep. The nicknames say it all: "fight or flight," then "rest and digest."

So, once again, you're looking for that happy medium. When your stress neurotransmitter levels are too high, you're in a panic, anxious and unable to settle down. When they're too low, you're unmotivated, listless, and unable to focus.

Furthermore, you want those biochemicals to remain high during the day, when you're working, focused, and meeting challenges, and then to decline in the evening and overnight, when you're relaxing and sleeping. Ideally, too, your stress neurotransmitters drop at every mealtime, so you can properly digest your food.

If your COMT is too slow, your stress neurotransmitters tend to remain in your body, causing you to be wired too much of the time. Conversely, if your COMT is too fast, your stress neurotransmitters leave your body too quickly and it's hard for you to build up enough "stress" to stay focused and motivated and on point.

Let's combine what we learned in the previous chapter with this new information. What happens if your body isn't moving through the Methylation Cycle efficiently? Well, then, as you may recall, you're going to be either short on SAMe or not able to utilize it very well. In either of those cases, your COMT isn't going to function optimally:

- If you have a slow COMT, it moves even more slowly, so that even more stress neurotransmitters and estrogen remain in your system longer than they should. Your dirty gene's weaknesses grow, while its strengths decline.

- If you have a fast COMT, being low in SAMe will at first *optimize* your mood and focus. You'll think, "Wow, I used to be so spacey, and now I get so much work done!" But if your Methylation Cycle is disrupted for too long, your fast COMT might start acting like a slow COMT. The stress neurotransmitters that used to fly out of your system are now staying in your system much longer than they should, and you feel wired, irritable, and overwhelmed by stress.

As you can see, your COMT requires a balancing act. Whatever type of COMT you were born with, you want to methylate the Goldilocks way: neither too fast nor too slowly, but just right.

Do *Not* Self-Medicate with SAMe!

I know that the Internet is full of SAMe success stories, making it seem like the miracle supplement of all time. And I'm not saying that you should never self-medicate with SAMe. If you learn how it affects you and when it's safe to take it, and if you really have a deficit, it can be a wonder supplement. But if you don't look at the big picture, SAMe can really mess you up.

One of my clients, a successful concert pianist, used to take SAMe before bed to help her sleep. Since she had a slow COMT, some extra methylation helped her get the stress neurotransmitters out of her system. But if she took SAMe on days when she *wasn't* stressed, she found herself tired, depressed, and crying all the time. On those days, speeding up her methylation got *too* many stress neurotransmitters out of her system!

I had another client who gave some SAMe to her "behavior-problem" kid—and his problems got even worse. When she put her son on the Clean Genes Protocol, he calmed down and became more cooperative, simply because he was now eating well and sleeping soundly and getting the exercise that his body craved, while staying away from the video games and sweets and chemicals that dirtied up his genes. She had found a lifelong approach to diet and lifestyle that brought out the best in her son.

You get the idea. *Please* don't run out to the store and buy supplements first thing. Put those car keys down and wait to complete your Soak and Scrub. If you do need supplements, you'll find out about it when you do your Spot Cleaning.

What Makes COMT Dirty?

Slow COMT

- Not enough SAMe
- Low homocysteine levels

- Excessive tea, coffee, and/or chocolate
- Too much stress, causing a buildup of stress neurotransmitters
- Excess weight or a diet high in animal fat, causing a buildup of estrogen
- Overexposure to xenoestrogens in plastics, personal-care products, or home and garden products, again causing a buildup of estrogen

Fast COMT
- Too much SAMe

Both Types of COMT
- Elevated homocysteine levels
- Insufficient vital nutrients, especially folate/B_9, cobalamin/B_{12}, and magnesium, all of which are crucial for both methylation and COMT
- A born-clean MTHFR that's acting dirty, or a born-dirty MTHFR that's not getting enough support

Dopamine Dangers

One of your COMT's main jobs is to methylate dopamine, converting it to norepinephrine. If your COMT is dirty—either born dirty or acting that way—you could end up with a form of dopamine known as *dopamine quinone*, which is very harmful to the brain. Dopamine quinone has been associated with medications used to treat Parkinson's disease and ADHD, as well as contributing to these diseases themselves, so you will want to consider your options carefully.

Even when your dopamine is in the form you want, you should avoid superhigh levels of it. Excess dopamine can make you agitated and irritable, and it can cause you to respond poorly to stress. The right amount of dopamine gives you an extra jolt that's just enough to get you performing at your best. But too much can cause you to freeze up, panic, and forget everything you ever learned.

I know actors and actresses, for example, who are terrific in front of the camera, where their moderate dopamine levels stimulate them to a wonderful performance. But in front of a live audience—talk about stress!—their dopamine levels go through the roof. As a result of that biochemical flood, they become paralyzed with stage fright, are overwhelmed by stress, and forget their lines. It's the difference between a fun roller-coaster ride and a truly terrifying trip down a mountain in a car with no brakes.

Each of us is unique, with our own personally ideal dopamine levels. If I tried to operate at Margo's dopamine levels, I'd probably burn out. If she tried to operate at mine, she'd be bored stiff. But even Margo can go overboard. Therefore, one of her tasks, as a "slow COMT person," is to learn how to take sufficient breaks and practice enough stress relief to keep her dopamine at manageable levels.

Likewise, Blake's calm, laid-back approach to life is a real strength, and he owes it to the naturally low levels of dopamine created by his fast COMT. But if Blake's dopamine levels get *too* low, he becomes distractible, unmotivated, and forgetful, and then he needs to find a way to rev himself up.

Both Blake and Margo need to discover the right balance of stimulation and relaxation that fits their genetic profile, their personality, and their health. You need to do that, too—we all do. The right balance may be different for each one of us, so we all need to find our own balance.

Levodopa

Levodopa is a dopamine-boosting medication commonly given to people with Parkinson's disease. This might seem to make sense, since Parkinson's is associated with low levels of dopamine.

As we've seen before, though, tinkering with any one part of your system can cause problems overall. When Levodopa raises dopamine levels, it puts a big strain on the COMT gene, thereby increasing dopamine quinone ... which in turn makes the Parkinson's worse. Parkinson's patients would do far better to live according to the principles of the Clean Genes Protocol and find gentler, more natural ways of boosting their dopamine levels—ways that improve function throughout their genes and neurotransmitters. I've had some success treating clients in this way, without the risks of medications such as Levodopa.

Ritalin and Adderall

If your kid struggles with ADHD, he or she might be prescribed methylphenidate (sold under the brand name Ritalin, among others), which increases dopamine. As you now understand, a fast COMT can lead to low dopamine levels, which are frequently responsible for poor focus and lack of motivation. So yes, bumping up the dopamine can sometimes help. As you can probably guess by now, though, the best way to do that is by starting, and sticking permanently with, the Clean Genes Protocol, which has helped many a kid around the globe.

Furthermore, certain drugs can create extra problems for a child with a fast COMT. Methylphenidate, for example, can turn a fast COMT into a slow COMT—and then your child struggles with *those* symptoms. To make matters much worse, methylphenidate may increase not only dopamine but also dopamine quinone, which, as we just saw, is toxic to the brain and may result in Parkinson's disease and other neurological disorders. The risks are just not worth it.

Now, what if you're an adult who thinks that Adderall—the so-called adult version of Ritalin—might help you focus and bear down?

Well, Adderall is basically an amphetamine. It's thought to make dopamine and norepinephrine more available to your brain. The problem is, if you use it too often, you get a rebound effect—your brain's levels of dopamine go down, waiting for that next hit of Adderall to bump them up. Overall, you get dopamine depletion, plus cell death from the amphetamine. And yes, Adderall can also generate dopamine quinone.

Your takeaway? The occasional use of Adderall will indeed stimulate your dopamine levels, and if you use it only once every few months, you won't notice any ongoing effects. Any more often, however, and you run the risk of long-term damage from dopamine quinone, especially if you have a dirty GST/GPX gene or if your body is struggling with an overload of heavy metals. There are better ways to stimulate dopamine production if you really need it—especially the Clean Genes Protocol.

ADHD Support Without Ritalin: Helping a Kid with a Fast COMT

Do these sound familiar?

- "Will you please stop moving incessantly while we're talking?!"

- "I asked you to take out the garbage. Every Tuesday night is garbage night. Why do I have to remind you each week?"

- "I swear, if your head weren't on your shoulders, you'd forget it."

- "Dad, I forgot my soccer jersey. Can you bring it? My game starts in ten minutes!"

Yep, that's me and my oldest son, Tasman. I love him and all my kids. But man, can they push my buttons!

Tasman is a fantastic kid. He's nearly a straight-A student, respectful (usually), and an incredible athlete. If you saw him in school or came over to the house for dinner, you would never know he struggles with ADHD.

He does, though. He has a fast COMT and a healthy Methylation Cycle, so he burns through his dopamine and norepinephrine like there's no tomorrow. To make matters worse, the kid is a kid. He's a tall, lanky, growing boy who doesn't eat nearly enough protein for all the soccer he plays and all the stress neurotransmitters he burns through.

So I'm constantly at him: "Will you please eat more protein so you can think clearly and bulk up?" I keep telling him that going to the gym to build muscle is a total waste of time if he doesn't eat adequate protein. Sheesh. Dads don't know anything, though, right?

So how does he manage to excel in school? How is he active and social and not bored and depressed? I know my biochemistry, and now so do you. You know that COMT's job is to burn through several biochemicals, including dopamine. The key, then, is to help Tasman, with his fast COMT, make more dopamine. And it turns out that dopamine is made from protein—specifically, from the amino acid known as tyrosine, which is found in both animal and vegetable forms of protein.

Can you guess what I give Tasman to further support his dopamine? That's right. Supplemental tyrosine. He has a bottle in his bathroom, and he takes a capsule every morning. I don't even have to remind him anymore. He's already told me, "Dad, I feel so much better when I take tyrosine that you never need to ask." Since he started building up his dopamine levels naturally, through protein and tyrosine, he's a different kid.

The system isn't foolproof, though. For about three weeks recently, Tasman was being irritable and just a punk. At first I was annoyed. Then I stopped being a dad for a moment and put on my doctor hat.

"Tasman, how many tyrosine capsules are you taking?"

He snapped, "Two, three, or four, depending on the day."

Bam. Because one capsule made him feel good, he assumed that multiple capsules would make him feel better. A reminder to me—which I sternly passed on to him—that dosages *matter*.

I told him to *stop* taking tyrosine until further notice, and within a day and a half I got my good kid back. Then he started forgetting stuff again and sleeping too much. I instructed him to take one tyrosine a day—but only when he feels he needs it. On days when he's got a lot to do—especially a lot of schoolwork or stuff that takes mental focus—he takes a tyrosine. On vacation, not so much. So that's where we are now.

"Tune in to your body," I tell Tasman. "Understand the biochemistry and how you're feeling." As he matures and starts increasing his protein intake, I'll likely reduce his tyrosine even further, especially as he becomes more self-aware and better able to judge what his body is telling him. Even when your self-medication is all through "natural means"—for example, the all-natural supplement tyrosine—cleaning up your genes is a balancing act, one that requires you to be continually aware of what you need.

Listen to your body and help your kids learn to listen to theirs. What you hear will always be better than what any doctor can ever tell you.

After all, your doctor is just a coach. You or your child is the athlete. It's up to you to get an awesome coach; but more important, it's up to you to put in the work—on and off the court.

Key Nutrients for a Healthy COMT

As you saw in chapter 5, the Methylation Cycle depends upon a number of nutrients: riboflavin/B_2, folate/B_9, cobalamin/B_{12}, protein, and magnesium. Because your COMT depends on the Methylation Cycle, it also depends on those nutrients.

That final nutrient, magnesium, is especially important for your COMT to function properly. So if you don't have enough magnesium in your diet—and about 50 percent of all U.S. residents don't—you're going to dirty up your COMT.

Magnesium: dark leafy greens, nuts, seeds, fish, beans, avocados, whole grains

Besides dietary insufficiency, there are two common reasons for magnesium deficiency: caffeine intake and the long-term use of a group of antacids called proton pump inhibitors (PPIs). When we get to the specifics of the Clean Genes Protocol, I'm going to help you stop using both caffeine and antacids while offering you some alternatives to promote better digestion and alertness. I don't drink caffeine or take any type of medication. None. You won't need to either.

No "Pill for an Ill"!

Remember, I don't want you running down to the vitamin store to buy magnesium or any other supplement—not until you've completed the first two weeks of your Clean Genes Protocol. You have some twenty thousand genes in your body. Let's not start medicating or supplementing them individually until we see what happens when we clean them all up together. If you try the supplement route first, you'll be frustrated. So don't.

Making the Most of COMT

I'll never forget the last conversations I had with Margo and Blake. Even though her COMT was slow and his—like Tasman's—was fast, they both said something similar:

Margo said, "I feel like knowing this biochemistry has helped me understand myself in a way I never could before. I always felt bad about being tense and wired and high-energy, like it was somehow my fault that I couldn't be like other people. Now I see that I just have a lot of extra dopamine. That's cool—that's very cool! But I can't let it get out of hand."

Blake said, "I always felt like I was maybe lazy and sort of slow. But it wasn't that—I just had low dopamine levels! I like the way I am, to be honest. But I'm glad I can do something about the parts of my biochemistry that don't always work so well."

For both Margo and Blake, the key was self-awareness. Margo needed to notice when her dopamine levels were climbing—when she was getting too tense, or too keyed up, or too wrapped up in work. She needed to make a conscious effort to take breaks, to slow down periodically, to balance her high-intensity action with some serious relaxation. She didn't need to "calm down" or "take life easy." She needed to figure out how to make her high-intensity style work better.

Blake needed to notice when he got spacey, forgetful, or unfocused. He needed to make a conscious effort to support his fast COMT with a high-protein diet, and maybe to give himself an occasional boost of tyrosine. He didn't need to push himself harder. Rather, he needed to learn how to "work smarter"—how to give his brain what it needed to complete the tasks he chose to undertake.

As noted above, the key to supporting your dirty COMT—whether it's fast or slow—is to be aware. You can't take appropriate action until you know what the problem is. Having been born with either a slow COMT or a fast COMT has both advantages and disadvantages, as we've seen, but you can't maximize the advantages unless you're self-aware. Now that you've had a little biochemistry lesson in this chapter, you have the tools to understand yourself better.

So I want you to do something right now. Yes, literally right now. Tune in. Pay attention to your mind and your body. Put the book down for a moment and ask yourself this question: How are you feeling right at this moment? Are you giddy? Excited? Irritated? Bored? Unable to focus? Depressed? Bothered by a headache? Which word or phrase describes you at this moment? With that information, how do you think your COMT is acting? Slow or fast? Do you think your COMT was born dirty or is just acting dirty? Does the feeling you've just identified in yourself come and go (suggesting that your COMT is acting dirty), or has it stuck around as long as you can remember (born dirty)?

My job is to give you the tools to understand how your genes are contributing to your mood and overall health. Your job is to give yourself the attention you deserve and act on what that attention reveals to you. And I promise: your moods have never experienced such consistent bliss as they will after four weeks on the Clean Genes Protocol.

Meanwhile, here are a few suggestions for making the most of your COMT. You can start these right away, even before you begin your Clean Genes Protocol:

For Both Slow and Fast COMT

- Optimize your weight, because body fat creates estrogen, making it harder for your COMT to regulate estrogen levels.

- Avoid as much as possible any contact between your food and plastic. Plastics are *xenoestrogens*, meaning that they mimic the effects of estrogen in your body. Your COMT is already struggling to optimize your estrogen levels. Why dump a lot of extra estrogen into your system?

- The subgroup of BPA plastics are also xenoestrogens, so avoid them, too—even if they're not in contact with your food. BPA can be found in a discouraging number of places—everywhere from the inside of cans to the outside of cash-register receipts—but do what you can to stay away from it.

- Meditate at least a few minutes every day. If you're overcharged, meditation will calm you down. If you're undercharged, it will help you focus.

- Go to bed and get up according to a fixed routine to help your body get the most refreshing sleep. If you're wired, a regular bedtime helps cue your body toward sleep. If you're scattered and unfocused, a regular bedtime helps create a routine that encourages focus. A sleep app such as Sleep Cycle can help. (See the Resources section.)

- Avoid the herbicide Roundup, as it affects aromatase activity. (*Aromatase* is an enzyme that converts other biochemicals to estrogen.) Also avoid all nonorganic soy and soy products, which have likely been farmed with Roundup. More generally, limit your exposure to all herbicides, pesticides, and other endocrine-disrupting chemicals in your home, garden, and personal-care products, including cosmetics. Especially damaging are glyphosate, phthalates, and dioxins.

- Eat as clean as you can. Buy organic produce, at least for the foods that are most likely to be exposed to industrial chemicals. The Environmental Working Group (www.ewg.org) has a list of foods—the worst offenders when it comes to toxins—that you should buy organic, and a list of foods that you can buy conventionally farmed.

- To balance your estrogen levels, eat more beets, carrots, onions, artichokes, and cruciferous vegetables (broccoli, cauliflower, kale, brussels sprouts, cabbage). Bitter vegetables such as dandelion greens and radishes support the liver, which metabolizes your estrogen, so load up on them as well.

- Be sure to eat a maximum of three times a day—balanced meals that each contain some protein, some carbs, and some fats. This way your blood sugar is balanced, and so are your moods.

- Declutter your home, office, garage, yard, and car. The more "noise" you have around you, the more "noise" you have in your head. That's the last thing you need! Keep things minimal and organized, and consider feng shui to order your environment.

For Slow COMT

- Monitor your stress levels throughout the day. Notice when you're getting more revved up or tense than you enjoy being. Develop ways to slow down, even for a minute or two—take a few deep, slow breaths;

listen to music; pause before each meal to appreciate the sights and scents of the food, so that when you start eating you're relaxed and not stressed.

- Make sure you get all the breaks, days off, and vacations you need. You might *feel* like a superhero—and most of the time you might even act like one—but overexertion is your kryptonite. Listen to your body and rest as needed.

- Burn it off. Exercise or sport of any type is a great way to burn through excess stress neurotransmitters.

- Be aware of how you feel in response to caffeine, chocolate, and tea. If you're feeling irritable or anxious from them, reduce your intake.

For Fast COMT

- Protein is your friend; sugar and refined flour are your enemies. Make sure to get high-quality protein at every meal—that is, organic protein that isn't deep-fried or buried in a white-bread sandwich. If you start the day with a starchy, low-protein breakfast, you're setting yourself up for a low-dopamine day, and your focus, motivation, and energy levels will suffer.

- Sleep is also your friend. During sleep your body makes more of everything you're short on—in the context of COMT, you need to give your body that time to make more dopamine. Everybody needs different amounts of sleep to function well. Figure out what *you* need and make sure to give yourself that advantage every day.

- Participate in brain-engaging activities like dancing, playing an instrument, sports, fast board games (not boring slow ones), and yes, even some video games. (Don't go overboard on the latter—they can be addictive and sometimes *too* stimulating.)

- Hugs! Hugs raise dopamine.

- While caffeine and chocolate may be helping you, don't overly rely on them: try the above suggestions instead. If you sleep well, eat right, participate in engaging activities, and get hugs, your need for stimulants from food and drink will go way down.

How the Clean Genes Protocol Supports Your COMT

Diet. Eating balanced meals up to three times a day will support your blood sugar and take some pressure off your COMT. (Constant snacking, by contrast, stresses your COMT and makes it even dirtier.) Eating only until you're 80 percent full also supports your COMT; eating until you're full, let alone stuffed, stresses all your genes. On the Clean Genes Protocol, you'll also get all the nutrients you need to support your Methylation Cycle and your COMT, because your diet will include lots of B vitamins and magnesium, as well as the right amount of protein—neither too much nor too little.

Chemicals. Avoiding industrial chemicals will lighten the burden on your liver, where estrogen is metabolized. Adding in sauna, hot yoga, Epsom salt baths, or any means of sweating is a fantastic way to support your liver and move toxins out of your body. We all have them—so we all need to detox.

Stress. Stress relief is a huge component of the Clean Genes Protocol. You'll learn to identify your stressors—news, social media, problematic friends, office work on weekends, news-laden TV, depressing movies—and you'll work to eliminate them. I'll also encourage you to identify what your hobbies are, including long-lost activities you absolutely loved when you were younger. Rekindle them and enjoy yourself! Your COMT will thank you for it. For example, I'm done writing this chapter, so I'm going kayaking with my wife! Follow my lead and reward yourself!

7

DAO: Oversensitivity to Foods

Hunter was a tall, quiet man in his forties who told me that he hated to complain. When I encouraged him to explain why he had sought my help, he began slowly, but I could hear his pain and frustration.

"I'm so tired of not knowing what I can or can't eat," he said finally. "One meal I'm good and the next I feel awful. I get headaches. I'm always irritable. If I eat the wrong thing, I get sweaty and my heart races. I have itchy skin and I get frequent nosebleeds. What the heck?"

I asked Hunter what his previous physicians had told him, and he shook his head.

"My wife finally convinced me to spend a ton of money on food allergy testing, and it showed nothing! We have a neighbor with allergies, and she told me to keep limiting my foods one by one until I found the culprit. But I don't know. It's just a never-ending battle."

When I confirmed that Hunter had several DAO SNPs, I understood the problem—and the solution. The problem was an oversensitivity to *histamine*, a biochemical that affects immune response and gut function. Histamine is also a neurotransmitter that affects thought and emotion.

Some people develop specific immune responses—either allergies or intolerances—to specific types of food. For those people, it may be possible to run a blood test and discover antibodies for responses to those specific foods. Another approach that works for some people is to do an elimination test—cut out all but a few "safe" foods from your diet and then add foods in one by one. When you have an unfavorable response, such as increased itching, headache, or faster pulse rate, you know that that particular food is challenging for you.

However, neither of these approaches worked for Hunter, because his problem wasn't about specific foods. His problem was the result of several factors working together:

- **A dirty DAO.** Hunter had been born with a lower-than-normal ability to process histamine. As a result, foods containing lots of histamine would likely—*but not necessarily*—be challenging for him.

- **A compromised Methylation Cycle.** If the DAO becomes overwhelmed, another gene takes over. That second, backup gene relies on methyl groups from SAMe. If the Methylation Cycle isn't working properly, those methyl groups aren't available.

- **Pathogens.** Any foreign pathogen triggers histamine release. Some pathogens simply cause histamine release, while others *create* histamine themselves. Identifying and eliminating gut pathogens is essential in overcoming a dirty DAO.

- **Food allergies.** If people eat a food they're allergic to, that allergen triggers histamine release and overwhelms the DAO. Oftentimes food allergies are caused by other problems, such as poor digestion or a leaky gut.

- **Leaky gut.** Leaky gut is a condition of the intestinal lining that allows partially digested matter into the bloodstream, where it triggers an immune response. (More on this later in the chapter.) When Hunter's gut wall was strong and had a lot of integrity, he could handle high-histamine foods fairly well. When his gut was leaky— which in his case was intermittent—histamine foods became more challenging. So a food that Hunter ate in, say, March, might give him trouble, while the same food eaten in, say, June, might cause no problems.

- **Poor digestion.** Again, poor digestion exacerbates the problem with a dirty DAO, while good digestion minimizes the problem. Weak digestion is defined as lower stomach acid, pancreatic enzymes, and/or bile. When any of these are low, it is easier for pathogens to invade the digestive tract and set up shop. This leads to higher histamine levels as the pathogens either trigger immune responses (which triggers histamine) or the pathogens release histamine themselves.

I laid out our master plan with Hunter. I told him that we were going to strengthen his gut, improve his digestion, and replenish his *microbiome,* the community of gut bacteria that are crucial for digestion as well as many other functions. During the initial phase, while his digestive system was improving, he should avoid leftovers—which increase in histamine the longer they sit (due to bacteria-producing histamine, which occurs even with refrigeration; freezing prevents this from happening, however)—as well as other high-histamine foods, such as cured meats, soured foods, dried fruits, citrus fruits, aged cheese (including goat cheese), many types of nuts, smoked fish, and certain species of fresh fish. Hunter could enjoy *some* of those foods—which he was relieved to hear—but I told him that we were going to find his sweet spot. As his gut healed and his microbiome became more robust, he would probably be able to increase the amount and variety of histamine-containing foods.

"Okay," Hunter said, when we had gone over our first steps. "I don't mind telling you, I'm relieved. It actually sounds like there's hope." He paused.

"I still wish I hadn't been born this way," he admitted. "With a problem like this, I mean. Other people seem to be able to eat what they want. I wish I could."

"I look at it just the opposite," I told him. "Other people can eat all sorts of food that isn't healthy for them, and for years they don't realize they *have* a problem. Maybe they're a little less energetic, or they get a couple of headaches now and then, or they have a little acne, or they get indigestion. But it all seems like small stuff, and they just brush it aside."

Hunter nodded.

"Then, when they turn forty, or fifty, or sixty, suddenly they have major problems—problems that have been building up for years and they didn't even notice. *Their* problems crept up on them. *Yours* are yelling loudly, trying to get your attention. You feel so bad from any unhealthy choice that you're really motivated to clean up your diet, get the sleep you need, and make all the other changes that your body is calling for."

Hunter stared at me for a moment. "I never thought of it like that," he said.

Your DAO in Action

Sad but true: pain and discomfort drive change, while comfort drives routine. I say that as somebody who has been struggling with symptoms and disorders created by my own dirty genes—including a MTHFR gene with only 30 percent function and a DAO gene that functions well below par. My whole childhood, I had major stomachaches. For most of my life, I wrestled with depression, irritability, chemical sensitivity, and a bunch of other symptoms. In college, I couldn't even enjoy a couple of beers with my teammates on the rowing crew. They could breeze through a campus party that would leave me hungover and miserable the next day. (I later learned that my vigorous exercise on the rowing team put a lot of demands on my Methylation Cycle, which meant that methylating alcohol was an extra challenge I just couldn't handle.)

But what has been the result of all those health issues? I now understand exactly what my body needs and which choices keep me feeling terrific. I lead an active life, working hard and playing hard—kayaking with my sons, hiking in the nearby woods, taking care of our huge garden, not to mention running my business, doing research, and now writing this book. I'm not sure I'd be in such good shape if my dirty genes hadn't forced me to take a long, hard look at what my body really needed.

So if you—like Hunter and me—were born with a dirty DAO gene, take heart. Your friends and loved ones born with clean DAO genes may be able to hold out longer as they damage their gut and microbiome, but *our* bodies are forcing us to make changes. In the long run, I think we're the lucky ones.

DAO: The Basics

Primary function of the DAO gene

The DAO gene produces the DAO enzyme, which is found in most organs but is especially plentiful in your small intestine, prostate, colon, kidney, and placenta (when you have one). The DAO enzyme helps process a key biochemical called *histamine.*

Your body's supply of histamine exists in two places: within your cells and outside your cells. Your DAO gene is focused on expelling the histamine that lives outside your cells, mainly in your gut.

- Some of that histamine is produced by the bacteria that live in certain foods, such as fermented foods, cured meat, and aged cheese.
- Certain probiotics, such as many species of *Lactobacillus,* produce histamine.
- Certain gut bacteria produce high amounts of histamine.
- Some histamine is produced by your own immune system in response to stress and to potential dangers from the foods you eat.

The right amount of histamine helps keep you healthy, but too much can overexcite your immune system, causing it to overreact to certain foods and even to your own tissues.

Effects of a dirty DAO

You tend to overreact to the histamine in your gut. As a result, you're more likely to develop food sensitivities and allergic reactions.

You might also *absorb* the histamine in your gut, which means it gets into your blood and then into your cells. When your cellular histamine is too high, you're vulnerable to neurological disorders such as Parkinson's disease.

Signs of a dirty DAO

Common signs include allergic reactions (such as hives, runny nose, and itchiness) and food sensitivities, car

sickness and seasickness, leaky gut, migraine, nausea/
indigestion, pregnancy complications, and SIBO.

Potential strengths of a dirty DAO

It's an advantage to be aware of allergens and trigger foods
right away, before they have the chance to make you sick.

Meet Your Dirty DAO

I know firsthand that a dirty DAO is no picnic. For as long as I can remember, I struggled with delayed symptoms after eating—symptoms that didn't show up for anywhere between twenty minutes and two hours after I had finished a meal. This time lag made it difficult to associate the symptoms with my food choices—especially since the symptoms were so varied. Maybe my pulse would start racing. Sometimes I'd get irritable, or hot, or my feet would start to sweat. (*What???*) I might develop patches of eczema on my neck, or maybe my nose would start to bleed. I might even have insomnia, unable to fall asleep with no idea what was keeping me up.

As you can imagine, I was miserable and frustrated. I was able to pinpoint a few problem foods—citrus fruits, wheat—and that helped a little. But not enough.

Years later, when I found out about dirty genes, I wasn't surprised to discover that I have a dirty DAO. I can enjoy *some* histamine-containing foods—I just can't overdo it. And now I know that if those pesky symptoms appear, they're probably the result of something I ate within two hours of when the symptoms showed up.

You've already run through Laundry List 1, so you've probably already figured out whether you have a dirty DAO. But here are a few more ways to track down this dirty gene:

☐ I'm often irritable, hot, or itchy after eating.

☐ I can't tolerate citrus, fish, wine, or cheese.

☐ If my skin gets scratched, it stays red for several minutes.

☐ I can't tolerate yogurt, sauerkraut, or kefir (a type of fermented milk).

☐ I can't tolerate shellfish.

☐ I can't tolerate alcohol, especially red wine.

☐ I can't tolerate chocolate.

☐ I have sweaty feet.

☐ I'm frequently itchy.

☐ I often get heartburn and frequently need an antacid.

☐ My eyes often itch.

☐ I have skin issues, such as eczema or urticaria (hives).

☐ I get frequent nosebleeds.

☐ I struggle with asthma or difficulty breathing.

☐ I get migraines or other headaches often.

☐ I get carsick, seasick, or generally feel dizzy.

☐ I have ringing in my ears at times, especially after eating.

☐ I seem to react to many foods.

☐ I've been told that I have leaky gut syndrome.

☐ I get diarrhea at times without any reason I can identify.

☐ I struggle with ulcerative colitis.

☐ I frequently have to take antihistamines.

☐ I frequently have a runny nose or am congested.

☐ I have trouble falling or staying asleep.

☐ I have blood pressure that's lower than 100/60.

☐ I struggle with asthma, exercise-induced asthma, or wheezing.

☐ I frequently have joint pain.

☐ I have arrhythmia.

☐ When I was pregnant, I could eat more foods than usual without symptoms.

☐ I get side effects from morphine, metformin, NSAIDs (medications like aspirin and ibuprofen), antacids, clonidine, isoniazid, pentamidine, and/or amiloride.

Health Conditions Related to a Dirty DAO

As we've seen, dirty genes can create health issues, whether those genes were born dirty or are simply acting dirty. Following are some of the disorders that researchers have associated with a dirty DAO.

- Anaphylaxis
- Arrhythmia
- Asthma/exercise-induced asthma
- Conjunctivitis or keratoconjunctivitis
- Duodenal ulcer
- Eczema
- Heartburn
- Insomnia
- Irritability
- Irritable bowel disorders, including colon adenomas, Crohn's disease, and ulcerative colitis
- Joint pain
- Nausea
- Parkinson's disease
- Pregnancy-related complications
- Psoriasis
- Vertigo

Meet Histamine: Your Problematic Ally

Like so many biochemicals, histamine is a double-edged sword. We depend on it—but it can cause a multitude of health problems. It all depends on *how much* histamine, *where* the histamine is located, and *how the rest of your body* is behaving.

One key function of histamine is to combat pathogens in your gut. After all, you never know what might be in your food or water. If a dangerous bacterium or a toxic substance is lurking there, histamine to the rescue! It stimulates your immune system to release killer chemicals that attack the dangerous invader and keep your body safe.

Histamine also plays a role in gut *motility*—that is, the ability of your intestines to keep first food and then waste moving through. You don't want either food or waste to linger inside you. The rotting material releases loads of toxins, which you want outside your body, not in. So thank you, histamine, for keeping things moving.

Finally, histamine helps your stomach secrete the acids it needs to digest protein. Once you've swallowed a bite of food, it lands in your stomach, where it breaks down into its components. Animal flesh, in particular, needs acid to break down fully, and histamine helps your stomach release enough acid to get the job done.

So you want histamine in your digestive tract—but not *too much* histamine. An excess of histamine can falsely trigger your immune system to release killer chemicals and create inflammation. Yet because, in this case of false alarm, there's no actual enemy to kill, your immune system is overstimulated for nothing and ends up hurting *you*—the very person it's trying to protect.

Enter DAO, whose job is to help your body clear out unwanted histamine. But if there's too much histamine, your DAO is overworked and can't do a great job. And if your DAO is not only overworked but also dirty, it doesn't function well even with ordinary amounts of histamine.

So, where does all that histamine come from?

Well, when you eat protein, you ingest a compound called *histidine*. Then, in the process of digestion, certain bacteria turn that histidine into histamine.

In addition, a lot of histamine comes from high-histamine foods— foods that *contain* live bacteria (fermented foods, including yogurt, raw sauerkraut, kimchi, and pickles); and foods that *were created by* live bacteria (aged cheese, cured meats). Fruit juice, alcohol, and kombucha are additional sources of histamine. Bone broth is all the rage these days, and boy, talk about a histamine bomb!

Your takeaway? Adding the histamine from these foods to the histamine that's already in your gut can be too much, especially if you have a dirty DAO.

High-Histamine Foods and Drinks

Here are some of the major histamine culprits:

- Aged cheeses
- Alcohol—all types, but especially champagne and red wine

- Bone broth
- Chocolate
- Citrus fruits and juices (except lemon, which is well tolerated by most)
- Cured meats: salami, some types of sausage, corned beef, pastrami, and the like
- Dried fruits
- Fermented foods, including yogurt, sour cream, kefir, raw sauerkraut, kimchi, pickles, and fermented vegetables
- Fish, especially smoked or canned; and certain types of fresh fish, especially when raw (as in sushi)
- Fruit juices
- Soured foods—for example, foods marinated in lemon or orange juice
- Tomatoes when raw; cooked are typically okay
- Spinach
- Vinegars (although some people do well with unfiltered, organic apple cider vinegar)

Your Marvelous Microbiome

Just a few years ago, almost no one had heard of the microbiome, and yet it's one of the most important parts of your anatomy.

Well, it's not exactly *your* anatomy. Your microbiome is composed of trillions of bacteria that live in your gut and elsewhere in your body, with cells outnumbering your human cells by a factor of 10 to 1, and genes outnumbering your human genes by a factor of 150 to 1. The microbiome evolved along with us, so there are many functions in our body that simply wouldn't work without the assistance of this microbial community.

For example, digestion. *We* don't digest fiber; our gut bacteria do. They ferment it as they feed on it, producing acids and other biochemicals that are vital to a number of different human functions, from digestion to the regulation of thought and mood.

You want a strong, diverse, robust microbiome that includes a variety of gut bacteria in the right proportions. Because when your gut bacteria go out of balance—when you have too much of some types and not enough of others—then, my friends, you've got trouble. Antibiotics, which kill dangerous bacteria but also destroy big swaths of your microbiome, can contribute to this imbalance. So can stress, a long-term illness or infection, poor diet, toxic exposure, and digestive issues such as leaky gut syndrome.

If your bacterial balance is disrupted by any of these factors, you may end up with excess histamine in your gut. As a result, your immune system will produce too many killer chemicals and a bunch of unpleasant symptoms.

Now, you might try to restore your bacterial balance by taking probiotics—powders, pills, or capsules containing live bacteria. Eating fermented foods—which, as we saw above, also contain live bacteria—is another great way to restore microbial balance. Often, this is a recommended approach, especially if you are or have recently been taking antibiotics.

But here's the rub. Fermented foods promote histamine, and so do some probiotics. On the other hand, some probiotics help your body to break down histamine. So in a perfect world, you want a balance: fermented foods and probiotics to support your microbiome; but overall, a healthy rather than an excessive level of histamine in your gut. This balance is harder to achieve—but all the more important—if you have a dirty DAO.

Your Biggest Defender

From your mouth all the way down to your anus is one long, continuous tube. Your mouth, throat, esophagus, stomach, small intestine, large intestine, rectum, and anus are all connected, with little valves between each portion of that digestive tract opening and closing as needed.

What happens when you eat and drink? The digestive process starts with your saliva, which allows your food and drink to slide down your throat, through your esophagus, and into your stomach, where they get processed by stomach acid. In the small intestine, your food and drink are further processed by digestive enzymes and bile. In the large intestine, beneficial bacteria do further processing. Whatever is left leaves your body as stool.

This long, continuous tube not only helps you digest your food but also protects you from harmful bacteria, parasites, viruses, and chemicals in your food and drink. This protection is carried out by built-in defenses: your stomach acid, digestive enzymes, bile, and microbiome. When any portion of this protection falters, your DAO is likely to be overwhelmed.

What Makes DAO Dirty?

- Too many histamine-containing foods.
- Too many histamine-containing liquids.
- Imbalanced microbiome.
- Small intestine bacterial overgrowth (SIBO).
- Disease or infection in the gut, caused by harmful bacteria, yeast (various *Candida* species), parasites, ulcerative colitis, Crohn's disease, and the like.
- Certain medications—antacids, antibiotics, metformin, and MAO inhibitors.
- An acidic diet.
- A high-protein diet.
- Gluten.
- Food sensitivities.
- Emotional/mental stress.
- Chemotherapy.

Hello, Leaky Gut

Amazingly, the lining of your gut wall is only one cell thick. Each of those cells fits tightly with its neighbors, held by so-called *tight junctions* (the barriers between cells designed to seal in any liquids, solids, or chemicals inside your gut). Within this wall—inside your gut—food is broken down into its most basic components—molecules of protein called amino acids; carbohydrates, such as glucose; fats, such as cholesterol; vitamins; and minerals. Only then are nutrients small enough to make it through those tight junctions. Everything else stays inside your digestive tract.

On the other side of your gut, your immune system stands guard in many places such as your bloodstream, liver, and spleen. This makes sense: if something sneaks out of your digestive tract and into the rest of your body, your immune system is standing by, ready to attack as needed.

Now, what happens if those tight junctions loosen up? This is a condition known as *intestinal permeability,* or *leaky gut.* Through these larger openings pass morsels of food that are only partially digested, a form that your immune system isn't equipped to recognize.

At first, these partially digested foods get through without being attacked. But if too many "invaders" cross your gut wall, your immune system fires up. It begins to tag these foods as dangerous invaders—a common response to cow's milk dairy, gluten (which also promotes the opening of the tight junctions), and a number of other foods.

Basically, whatever you eat frequently can trigger an immune reaction when you have leaky gut syndrome. This is why people often have trouble with foods they used to be able to eat. They remove those problem foods, introduce new ones, and then a month later they're reacting to the new foods too! Sound familiar?

Your immune system creates antibodies geared to recognize any food that leaks through. Once they've been created, whenever you eat even a tiny bite of the problem food, the antibodies cue your immune system to send out killer chemicals. The more often you eat these foods, the harder your immune system will work to deal with the war you've created. Your joints hurt, brain fog settles in, and you're tired.

This immune response is also known as *inflammation.* As we've seen, *chronic inflammation*—the sort of inflammation that never goes away because it's continually triggered and retriggered by diet, stress, and other chronic factors—is bad for your health.

And what else is going on while your gut is leaking? You guessed it: your body is generating extra histamine. The histamine is intended to calm the inflammatory process, but too much of it creates a vicious cycle, retriggering the immune system and causing the release of yet more histamine. All of this makes it hard for your leaky gut to heal, as well as dirtying up your DAO. So now you've got at least three vicious cycles (gut, immune system, histamine), each of which is making the other two worse—and further burdening your DAO.

To make matters even worse, the DAO enzyme—whose job, we've noted, is to process histamine—lives in the cells of your gut wall. So if your gut wall is distressed, with fewer cells and less integrity, you're going to have less of that DAO enzyme—and thus even fewer resources for processing histamine. This is why healing a leaky gut can vastly increase your tolerance for certain foods. Finally, with tight junctions, you have the DAO enzyme to process those foods.

Luckily, by paying attention to diet, exercise, sleep, toxic exposure, and stress, you can replenish your microbiome, heal your leaky gut, and lower your histamine levels. All of those steps lighten the burden on your DAO, replenish your DAO enzyme, and ensure that you have enough methyl groups to keep your backup histamine gene well methylated.

What About Antihistamines?

Lots of people ask me whether antihistamines are effective against the problems addressed in this chapter. It's a logical question. If your DAO is overwhelmed by *too much histamine,* you might think you could support it by taking an *antihistamine* (a compound that blocks the body's response to histamine).

There are two ways to answer this question. One is, "Maybe. Depending on which antihistamine you take, your symptoms might reduce or disappear altogether."

Zyrtec, for example, a popular antihistamine used for seasonal allergies, impacts how histamine binds to histamine receptors, thus mitigating your symptoms.

Likewise, Benadryl blocks your histamine receptors so that histamine can't bind to them. Since mania and insomnia are both associated with high histamine levels, some doctors will even prescribe Benadryl for those conditions. And indeed, the symptoms reduce.

Notice, though, that I didn't say Zyrtec or Benadryl *lowers* your histamine levels. You've still got high levels of histamine. They're just not binding to the receptors that normally receive them. The second you stop taking the medication, your histamine levels bind where they normally do, and your symptoms come *roaring* back. This yo-yo effect keeps you reliant upon the antihistamine drugs.

So here's my second answer, which I like better: "Do you want to be on antihistamines for life, or do you want to solve the problem at its root?" Because the way to solve the root cause of your problem is—you already know this!—to clean up your genes. It's not always easy. But it *is* simple!

The Antacid Connection

Here's another problem with too much histamine in your gut: it causes acid reflux and heartburn. In fact, the class of antacids known as *proton pump inhibitors* (PPIs) behave like antihistamines and block your histamine receptors.

But, like antihistamines, antacids don't lower your histamine levels; they just change your body's response to histamine. I would so much rather you stopped eating high-histamine foods and cleaned up your DAO and other dirty genes. That's a much better long-term solution than being hooked for life on Prilosec or Zantac.

Key Nutrients for a Healthy DAO

The two primary nutrients that your DAO gene needs to work properly are calcium and copper:

Calcium: kale, broccoli, watercress, sprouted grains and beans, lower-histamine cheese (again, goat or sheep), bok choy, okra, almonds

Copper: beef liver, sunflower seeds, lentils, almonds, blackstrap molasses, asparagus, turnip greens

You also need to nourish your body with foods that balance high-acid or acid-generating foods:

Almond milk
Artichokes
Arugula
Asparagus
Avocado oil
Beets
Bok choy
Broccoli
Brussels sprouts
Buckwheat
Cabbage
Carrots
Cauliflower
Celery
Chia seeds
Coconut
Coconut oil
Endive
Flax
Garlic
Ginger
Goat's milk
Grasses—such as wheatgrass, barley grass, alfalfa, and oatgrass
Green beans
Himalayan salt
Kale
Kelp
Leeks
Lentils
Mustard greens
Okra
Onion
Peas
Quinoa
Rhubarb
Sea vegetables
Sprouts of any type
Watercress
Zucchini

Making the Most of DAO

Hunter and I spoke a few weeks after our first meeting, and he was doing much better. He had already started to figure out which histamine-containing foods gave him the most trouble (in his case, sauerkraut, pickles, salami, and red wine) and which he could sometimes manage in small amounts (aged goat's or sheep's milk cheese, yogurt, and kefir). In addition, he was making good use of a supplemental probiotic that helps process histamine.

Hunter also realized that he needed to limit leftovers, focusing only on fresh food. (Remember, the older the foods, the more histamine-producing bacteria are present.)

"It's sort of a pain," Hunter told me in his quiet way, "but it's worth it, because all those symptoms are gone and I've got more energy than I've had in years." Although Hunter had cut out fermented vegetables and raw sauerkraut, those foods have enormous benefits for the microbiome. I reassured him that, as his system got stronger, he'd be able to add some back in.

Here are some other suggestions to support your DAO. You can start them right away, without waiting to begin your full Soak and Scrub:

- Stop taking probiotics containing *Lactobacillus casei* and *Lactobacillus bulgaricus*. (You have to read the labels closely to discover specific ingredients.) In chapter 15, you'll get some tips on which probiotics to take.

- *For women:* Check your estrogen levels, especially if your histamine symptoms get worse around the time of ovulation—ten to fourteen days after your period ends. High estrogen levels can trigger your body to release more histamine. Be sure to follow the estrogen-balancing suggestions in chapter 6: avoid plastics; optimize your weight; and eat more beets, carrots, onions, artichokes, dandelion greens, radishes, and cruciferous vegetables (broccoli, cauliflower, kale, brussels sprouts, and cabbage).

- Support your digestion so that you have adequate stomach acid, digestive enzymes, and bile flow. These are all essential to keep your microbiome healthy and pathogens out. (I'll go into detail about how to do this in my discussion of the Clean Genes Protocol.)

- Counter foods that *generate* acids with foods that help *reduce* acids. Make sure your meals are balanced. For example, if you're eating a lot of protein, have a lot of steamed vegetables to go with it. If you have a little bit of kombucha, have some sprouted greens to go with it. Use the foods from the list on page 130 to balance foods from the list on pages 123 and 124.

- Optimize sleep and reduce stressors, because stress neurotransmitters increase histamine release. Effective sleep supports include meditating, using blue-light filters on computers and other screens, avoiding screens one hour before bedtime, sleeping in a dark room or in a good eye mask, and monitoring your sleep with such apps as Sleep Cycle or ŌURA.

How the Clean Genes Protocol Supports Your DAO

Diet. We'll make sure you reduce your consumption of high-histamine foods and drinks. We'll ensure that your DAO enzyme can function optimally by providing it copper and calcium. We'll also have you balance your meals so that your body's acid levels are low enough to allow your DAO enzyme to work.

Chemicals. In a variety of ways, we'll support your digestion completely so that you avoid pathogenic bacteria coming in and taking hold in your gut. Adequate stomach acid, pancreatic enzymes, and bile are essential here—and the diet and supplements on the Clean Genes Protocol ensure that you get them.

Stress. Stress neurotransmitters limit your ability to release your stomach acid, digestive enzymes, and bile. On the Clean Genes Protocol, you'll practice stress reduction and stress relief, calming you down so that your body can digest your food and keep bad bacteria out.

8

MAOA: Mood Swings and Carb Cravings

Just as there are two COMT profiles—fast and slow—so are there are two MAOA profiles, also fast and slow.

Keisha had a fast MAOA. She craved carbs and chocolate like there was no tomorrow. At our first session, she was at least sixty pounds over her ideal weight and extremely frustrated with what she saw as her own weakness.

"I *know* better, but I just can't seem to help myself," she told me. "It's like I have no willpower. I stock up my fridge with healthy vegetables and I make a nice healthy dinner with broiled chicken and greens and maybe a salad. And then, after an hour or two, I crack. I have to have some carbs. If I don't run out to the corner store and get myself a candy bar and a couple of little cakes, I feel like I'll sink into a black hole. Some nights I resist and some nights I don't, but either way, it's driving me crazy. I feel like the weakest person on earth, and I'm beginning to hate myself for it."

"I look at it completely differently," I told her. "You're not a weak person—far from it! You're fighting hard to listen to your body and give your body what it needs. The only problem is, you don't have the right tools to do that—yet. And we're going to fix that today."

Keisha had sent me her genetic testing results, so I already knew that she had a fast MAOA. Even without the results, though, I would have been pretty sure: she fit the profile perfectly. I said, "Let me ask you something. How do you feel *before* you eat the carbs or chocolate?"

Keisha frowned. "Before, it's like I'm going deeper and deeper into a dark hole. I know that I'll feel better after eating the carbs, but I also know that I shouldn't. There's this constant guilt going on, but the blue feeling is too much to handle, so I always crack."

I nodded. "And after?"

Keisha shook her head. "After, I *do* feel better. I feel more hopeful—more at peace." She laughed. "For about one hour, tops. And then I crash again. The cycle just keeps repeating, and I'm so tired of it. That's why I'm talking with you. It has to stop."

I explained to Keisha that her dirty MAOA gene caused her body to process serotonin very quickly. Serotonin is a neurotransmitter that helps us feel at peace, optimistic, and self-confident. When our serotonin levels are low, we feel depressed, bleak, and helpless, without much faith in ourselves or our abilities.

Thanks to her dirty fast MAOA gene, Keisha's serotonin levels fell way too fast. She knew only one way to pick them back up again—with sweet, starchy foods that *do* offer a temporary serotonin boost, probably the fastest one around. So yes, Keisha's system worked quickly and was definitely effective.

But there were negatives that she knew all too well: emotional roller-coaster rides and weight gain. "You're actually on the right track," I told Keisha. "You've just missed the mark a bit. This isn't about willpower or gluttony or any of the other shame words we like to use. Erase those right out of your mind, because we're not talking about willpower. We're talking about basic biology—the human body you were born with."

I asked Keisha one more question, one that I knew would surprise her. "So—do you find yourself often waking up in the middle of the night and needing a midnight snack in order to fall back asleep?"

She slammed her hand on the desk and said, "Yes! How did you know that?"

I explained to her that one of the things serotonin is needed for is to make melatonin, a hormone key to good sleep. Since she was burning through serotonin so fast, her melatonin levels were low. She immediately asked me how much supplemental melatonin to take.

"Let's start with the foundations," I told her. "The secret isn't to take another pill, but to give your body a loud, clear message that it's got plenty of food. Research has shown that if you consume a moderate amount of protein throughout the day, your neurotransmitters will be more stable. This will reduce your food cravings, mood swings, and tendency to binge, and it will also help you sleep through the night."

Keisha admitted she was more likely to hit the road running with a donut and coffee than with eggs or other protein. She was thrilled to imagine that switching her morning routine might break the cycle.

Now, there's one more factor to consider: nutrients. Your MAOA gene needs a steady supply of riboflavin/vitamin B_2 to work its magic. It also needs a nutrient known as *tryptophan,* which is found in carbohydrates. Keisha's fast MAOA was burning through both at a rapid rate, so naturally she craved the foods that would quickly replenish them—that is, carbs. I told you she wasn't far off the mark! The only problem was that she chose the wrong carbohydrates and didn't balance them with the right other foods. She needed to choose carbs that would affect her body more slowly—complex carbs that include plenty of fiber—and to balance them with protein and healthy fats. That diet would calm her fast MAOA down to a normal rate. No more cravings. No more roller-coaster rides. And no more weight gain.

Finally, I talked with Keisha about different ways to manage her stress. People often think stress relief is a touchy-feely extra, something separate from the hard business of science and biochemistry.

Nope! Stress is one of the most powerful biochemical experiences your body knows. And managing stress is one of the best things you can do for your genes, your overall health—and yourself. When you're calm, your genes work one way, and when you're stressed, they work another. That's why stress reduction belongs at the very top of your list.

After Keisha's appointment was over, I met with Marcus, whose profile suggested that he too had been born with a dirty MAOA—one that was too slow. As a result, he often struggled with a hair-trigger temper, going from relatively calm to absolutely nuts in a matter of seconds. He often found himself jumpy, anxious, and easily startled.

From Marcus's point of view, getting mad wasn't the problem. "Look, if you were me and these things happened to you, you'd be mad, too," he said frankly. What concerned him was that once he did get irritated, it took him hours to calm down. Sometimes, he couldn't calm down at all, and his whole day was shot.

Over and above the anger itself, Marcus didn't like feeling out of control. "I don't like myself this way," he told me, "and it's really taking a toll."

As with Keisha, the problem wasn't in Marcus's willpower or mental control. It was in his genes. His dirty slow MAOA moved stress neurotransmitters out of his body more slowly than normal. If something upset him, he'd feel a spike of dopamine and norepinephrine, as we all do. But a clean MAOA gene helps move those stress neurotransmitters out of our body quickly—we get mad, or stressed, or excited, and then we get over it. Marcus's slow MAOA was moving those neurotransmitters out of his body way too slowly. That made it difficult for him to calm down, just as Margo, who with her slow COMT was profiled in chapter 6, had a hard time powering down after a demanding day at work.

There are some upsides to having a slow MAOA. The high levels of dopamine and norepinephrine in Marcus's brain meant that he was always ready to take on any challenge he needed to. But those high levels also kept him cranky and irritable, with his temper often out of his control. His biochemistry wasn't allowing him to calm down.

When I explained all this to Marcus, I could see the waves of relief pass through him. "Finally. I knew it! That's exactly how it feels—like something *won't let me* calm down. This means the world to me, Doc. Now, what can we do about it?"

I told him I wasn't done yet. To make matters worse for Marcus, I explained, his MAOA was processing serotonin far too slowly.

At first glance, you would think that was a good thing, right? You would imagine that Marcus's serotonin levels were always high, so he felt optimistic, calm, and self-confident all the time.

I'm sorry to tell you it doesn't work that way. Yes, low serotonin makes you feel bleak and unconfident, as Keisha so often felt. But *high* serotonin makes you feel anxious and irritable. Just as Keisha's fast MAOA was leaving too little serotonin in her system, Marcus's slow MAOA was leaving too much serotonin in his.

As with Keisha, diet was the first step. Marcus also needed a solid breakfast full of balanced protein, and he needed to limit carbs and sugars throughout the day. These "high-energy" foods tend to cause energy levels to spike, then crash. Then stress neurotransmitters follow suit, and the roller-coaster ride begins. Since Marcus's stress neurotransmitters were already being moved out of his body too slowly, he didn't need a sugar rush or a carb high to exacerbate the problem.

Snacking—something Marcus confessed to—would also tend to imbalance his blood sugar and therefore his stress neurotransmitters. I advised him to stop eating between meals and instead get all his healthy foods in at breakfast, lunch, and dinner.

I also wanted Marcus to work on stress relief, to reduce the work that his MAOA had to do. After all, the more stressed you are, the more dopamine and norepinephrine your brain releases.

Finally, I told Marcus that when he's stressed, getting solid sleep is even more of a requirement. Without sleep, we become more emotionally reactive. I wanted him to be asleep by 11 P.M. and to limit electronics before bed. Those bright screens disrupt the body's production of melatonin—the blue light fools the brain into thinking it's daylight. And Marcus's slow MAOA meant that ample stress neurotransmitters were already keeping him awake. If we could just help him calm down and fall asleep, he'd be golden, as that excess serotonin could become melatonin, which would keep him asleep.

Your MAOA in Action

Okay, confession time. I have a dirty MAOA, and boy, does it show! I'm a sucker for carbohydrates—especially sugar. I used to eat a half gallon of ice cream in one sitting. Did I just admit that? Yes. And it's true!

So which dirty MAOA do you think I have? Fast or slow?

I crave carbohydrates. That's your clue. My MAOA is fast, like Keisha's, which predisposes me to carb bingeing.

That personal history is how I know that Keisha—and you—can overcome the effects of a dirty MAOA, and the remedy has nothing to do with willpower. It's all about balance. When I revolutionized my eating—started the day with a high-protein breakfast and made sure to get adequate protein throughout the day—my cravings began to subside. I realized that I had to go cold turkey on sugar, because otherwise my cravings would cycle back up. But eating adequate protein made that transition a lot easier.

When I get stressed, I do feel those cravings again. But now I have the tools to handle them. It's up to me to recognize what's happening and stop pushing myself so hard, or find a way to bring my stress levels down—for example, by going for a walk or meditating. When I do, my cravings go away again.

For people with a slow MAOA, the story is similar. Eat the right amount of protein, reduce sugar intake, and monitor your stress levels. It's amazing what a huge difference those three steps can make. In addition, those with a slow MAOA may need a lighter dinner in the evening so that they don't go to sleep with lots of stress neurotransmitters in their system.

MAOA: The Basics

Primary function of the MAOA gene

The MAOA gene produces the MAOA enzyme, which helps you process two key stress neurotransmitters—dopamine and norepinephrine—both of which enable you to rev your body up for the stress response. MAOA also helps you process *serotonin,* a neurotransmitter that enables you to feel calm and optimistic.

Effects of a dirty MAOA

A dirty MAOA sets you up for tremendous mood swings, especially if you were also born with a dirty MTHFR and/ or COMT. The combination of these three born-dirty genes can give you tremendous energy and focus—but can also make it hard for you to control your temper or rise above irritating situations.

Slow MAOA. A slow MAOA eliminates norepinephrine, dopamine, and serotonin more slowly than usual, which can set you up for an excess of these neurotransmitters.

Fast MAOA. A fast MAOA eliminates norepinephrine, dopamine, and serotonin *too* quickly, which can set you up for a shortage of these vital neurotransmitters.

Signs of a dirty MAOA

Slow MAOA. Common signs include difficulty falling asleep, overactive startle reflex, headaches, irritability, mood swings, prolonged anxiety, rage and/or aggressive behavior, and trouble relaxing or powering down.

Fast MAOA. Common signs include alcoholism and/or other addictions, ADHD, carb and sugar cravings, depression, difficulty staying asleep, fatigue, and flat affect.

Potential strengths of a dirty MAOA

Slow MAOA. When you're not stressed, you can be more alert, attentive, cheerful, energetic, focused, productive, and self-confident.

Fast MAOA. When you're stressed, you have a greater ability to calm down. You're generally more relaxed and easygoing.

Meet Your Dirty MAOA

Completing Laundry List 1 has given you a good idea of whether you have a dirty MAOA. But if you'd like some more specific questions to round out the portrait, here you go:

- ☐ I've been diagnosed with ADHD.
- ☐ Major depression is common in my family.
- ☐ Alcoholism is present in my family.
- ☐ I feel that I'm addicted to carbs.
- ☐ I do better when I eat more protein.
- ☐ I find that I breathe faster when I'm stressed.
- ☐ I tend to become aggressive more often than I'd like.
- ☐ It often takes me a while to slow down.
- ☐ I can focus for a long time.

Health Conditions Related to a Dirty MAOA

Whether your MAOA is fast or slow, it can create a number of health problems for you. Note how many conditions on the following list are neurological or mood disorders. That's because your MAOA is involved in processing *neurotransmitters*, the biochemicals that enable communication within and from your brain.

- Addictions, such as alcohol, nicotine
- ADHD
- Alzheimer's disease
- Antisocial personality disorder
- Anxiety
- Autism
- Bipolar disorder
- Depression
- Fibromyalgia
- Irritable bowel syndrome
- Migraine
- Obsessive-compulsive disorder
- Panic disorder
- Parkinson's disease
- Schizophrenia
- Seasonal affective disorder

Going Gray: Part 1

Ever see a person who was under tremendous stress whose hair turned gray or white in a matter of days? That's a real thing, and it's caused by hydrogen peroxide—not applied from the outside, but produced from within.

During times of stress, your body produces a lot of norepinephrine and dopamine, as we've seen. Your MAOA needs to remove them from your system—and a natural by-product of that process is hydrogen peroxide. That by-product is in turn removed by glutathione, your body's prime detox compound.

If extreme stress continues, your MAOA keeps eliminating your excess stress neurotransmitters as fast as it can—it works triple-time if need be—all the while producing way too much hydrogen peroxide. Glutathione tries to keep up, but it can't; it's not an easy compound to make, and eventually your body runs out of it. In the end, the excess hydrogen peroxide wins out and discolors your hair, turning it gray or, if the intense stress persists, eventually white.

I wish this were just a cosmetic issue. It's not. Too much hydrogen peroxide affects more than hair color. It's bad for your brain, too, commonly resulting in behavioral issues—erratic moods, memory problems, irritability, and aggression. It can even lead to neurological problems such as amyotrophic lateral sclerosis (ALS), Parkinson's disease, or Alzheimer's disease.

In other words, stress is serious. Figuring out how to reduce it and relieve it is just as important to your health as getting all your vitamins or reducing your exposure to toxins, *especially* if you have a dirty MAOA.

The Tryptophan Steal

Why is it that when you're stressed, you crave carbohydrates and sugar? There are a number of reasons, but let's focus on the one that relates to MAOA.

As we discussed earlier, your MAOA enzyme produces serotonin. To do so, it requires tryptophan, which comes largely from carbohydrates. Yes, there's some in protein as well, but the tryptophan in protein does not cross into the brain easily. People often blame the tryptophan in a holiday turkey for the postmeal lethargy they feel, but that's a myth. Most of your Thanksgiving tryptophan is actually from the carbohydrates you eat—such as sweet potatoes and stuffing—and yes, it *does* put you to sleep!

Here's what makes tryptophan so complicated: there are two places it can go. The calmer you are and the less your body suffers from inflammation, the more of your tryptophan goes toward making serotonin. The more stressed out or inflamed you are, the more of your tryptophan goes toward making quinolinic acid, a substance that's bad for your brain. In other words, stress steals your tryptophan.

This is why, if you're under a lot of stress or are dealing with chronic inflammation or have a chronic disease—which by definition is stressful and inflammatory—you crave carbohydrates. Your tryptophan is being stolen so fast that your MAOA *acts* dirty even if it wasn't born that way.

Remember, it's not just the tryptophan. As your tryptophan levels drop, so do your serotonin levels. Suddenly you're depressed, and you find yourself bingeing on chocolate and carbs just as Keisha did. What's more, with your serotonin levels tumbling, you don't have enough serotonin to make melatonin, so now you can't sleep. Sound familiar?

The key to keeping your tryptophan from being stolen is to identify your stressors and reduce the inflammation that creates and accompanies chronic disease.

What Makes MAOA Dirty?

Slow MAOA
- Too much tryptophan
- Too little riboflavin/vitamin B_2

Fast MAOA
- Too little tryptophan
- Too much riboflavin/vitamin B_2

Both Types of MAOA
- Too little glutathione
- Chronic stress
  - Physical stress, such as blood sugar imbalance, infection, yeast overgrowth, SIBO, leaky gut, or anything else that puts an ongoing physical strain on your body—including improper breathing (for example, holding your breath while concentrating; or breathing shallowly, from your chest, instead of deeply, from your abdomen)

— Emotional stress, such as demands at work, home, or in your personal life that create an ongoing emotional strain

Chronic inflammation

- From your diet—excessive eating, or eating foods that you react to with allergies or intolerance
- From chronic physical or emotional stress
- From chronic conditions such as obesity/overweight, cardiovascular disease, diabetes, autoimmune conditions, and cancer—which both create inflammation and are made worse by it

---

MAO Inhibitors: Mere Band-Aids

Millions of people in the United States are depressed. Billions of dollars are spent annually trying to identify a drug that can reverse depression. I'll tell you right now: depression is a complex disorder, and while research has implicated an imbalance in neurotransmitters, the blame doesn't lie with a *single* neurotransmitter. Drug companies have researched the MAOA enzyme thousands of times. They've made drugs that slow down this gene in order to keep serotonin in the brain longer, the goal being to help people recover from depression.

How's that working? For most people, it's not. The real problems behind depression are inflammation and stress. Depression isn't a serotonin deficiency. It's a health deficiency. Chronic disease drives depression.

With the Clean Genes Protocol, we will reduce your stress and inflammation significantly. After four weeks, your moods should be moving in the direction you want—and the direction you deserve.

Of course, do *not* stop taking an MAO inhibitor or any other type of antidepressant or anti-anxiety medication without your doctor's supervision. You could set yourself up for major problems if you stop cold turkey or taper off on your own.

Key Nutrients for a Healthy MAOA

For your MAOA to function properly, you need two compounds: riboflavin and tryptophan.

Riboflavin/B$_2$: liver, lamb, mushrooms, spinach, almonds, wild salmon, eggs

Tryptophan: spinach, seaweed, mushrooms, pumpkin seeds, turnip greens, red lettuce, asparagus

Once again, I suggest that you load your diet up with these foods, rather than taking supplements. Your body is always happier with fresh, whole foods than with any other form of nutrition. Remember, while tryptophan is found in protein, it isn't as well absorbed as the tryptophan you get from carbohydrates.

Making the Most of MAOA

Remember how in chapter 6, Margo and Blake realized that self-awareness was their key to managing a dirty COMT?

Well, that same self-awareness is your key to managing a dirty MAOA. Whether your MAOA is fast or slow, you need to notice certain warning signs, which are your gene's way of asking you to slow down and give it some more support.

Each of us has our own warning signs, although it might take some work for us to recognize them. I asked Keisha and Marcus to identify their warning signs, and here are their lists. Are their warning signs anything like yours—or would you make a different list?

Keisha's Warning Signs (FAST MAOA)

- I've gotta have that chocolate!
- I'm dreaming about that sugar!
- I'm starting to feel blue again.

- I woke up again in the middle of the night and needed a snack to fall back asleep. So tired of this. I just want to fall asleep and *stay* asleep!

Marcus's Warning Signs (SLOW MAOA)

- I'm staring at the ceiling again, not able to fall asleep.

- I find myself irritable about nothing—the kids joking around, my wife on the phone when I get home, little things. If I start getting annoyed at nothing, that's a bad sign.

- I can't calm down. That tells me I've let things go too far and I've messed up somehow by not eating right, not sleeping enough, or piling on the stress.

- I've got a headache again, one that's already pretty far along. I'm looking to recognize the warning signs *before* a headache.

- I'm holding my breath frequently—like when I buckle down to concentrate or work hard—or I'm breathing more shallowly.

As you build your self-awareness, you'll start to discover things you can do to interrupt a stress pattern and stop your cravings.

For Keisha, it helped to eat protein with every meal and have some high-protein snacks in the fridge at work—pumpkin seeds, a slice or two of turkey, some hummus and carrots. I didn't want her to snack too often, but if she was going to snack, protein was better than something sweet or starchy. At dinner, she learned to have some tryptophan-rich foods to provide her with much-needed serotonin and melatonin.

As Keisha cleaned up her genes, she also found herself losing weight—without even really trying—for the first time *ever*. Until she went on the Clean Genes Protocol, Keisha had always experienced weight loss as a huge effort requiring massive amounts of willpower—only to find herself failing as she caved a month later.

"Since I started adopting these changes, my cravings have gone way down. But I didn't notice it at first; it just happened. Then a friend of mine at work told me, 'You seem so much more upbeat and—hey, you've lost some weight, too!' That was a nice surprise."

How did this happen? First, Keisha was having protein with each meal and not waiting until she felt starving to eat. This led to her no longer craving carbs and sugar, which meant that her blood sugar was stable and her metabolism was working efficiently; this also improved her mood. Finally, her cells were getting what they needed to burn fuel, which also improved her metabolism. Keisha loved feeling full and satisfied without overeating, and she was thrilled to see her weight optimize—without a fight!

Learning about the Tryptophan Steal convinced Keisha to practice some stress-release techniques, including deep breathing, listening to music, and leaving the scene. (Saying "Excuse me, I need a bathroom break" when things get tough is a good stress-buster!)

Marcus also relied on leaving the scene when he found himself getting irritated, especially at home. Now that he'd learned he was genetically susceptible to remaining irritated, he knew he had to manage his anger. He'd step outside for a brisk five-minute walk or go to another room and look out the window, which helped him recharge and reset. He also paid deliberate attention to his mood, his breathing, and his body.

"I've figured out that when I'm stressed at work, I have to be especially careful about my diet and my breathing," he told me. "When I'm on vacation, I'm much more relaxed, so I can ease up a bit with my diet and still be fine."

How the Clean Genes Protocol Supports Your MAOA

Diet. Eating balanced meals helps keep your neurotransmitters in balance. Making sure to have protein, healthy carbohydrates, and healthy fats *every time you eat* is key. Don't say, "I had a high-protein lunch, so I can just have some brown rice for dinner." You need to balance *every* meal. That doesn't mean the proportions always have to be identical, however. For example, one meal you might have more protein, fewer carbs, and a touch of fat, while the next meal might be low-protein and low-carb with a bit more fat. Limiting sugar and processed foods and making sure not to overeat are musts; otherwise, you mess up your blood sugar and trigger mood irregularities.

Chemicals. Your MAOA produces a ton of hydrogen peroxide when you're stressed out, as we've seen. This depletes your glutathione, which means you have far less protecting you against heavy metals and chemicals. On the Clean Genes Protocol, we'll help you avoid stress so that you can conserve your glutathione.

Stress. The Tryptophan Steal is a real deal. Don't fall victim to it. By implementing stress reduction techniques, you can turn your tryptophan into feel-good serotonin and sleep-well melatonin, instead of sending it down the road to make brain-harming quinolinic acid. You need to figure out which stress-reducing techniques are right for *you*. As for me, now that I'm done with this chapter, I'm off to go hiking in the forest with my wife and boys.

9

GST/GPX: Detox Dilemmas

When I had my first meeting with Megan, she was at her wit's end.

"Everyone in my family thinks I'm too sensitive," she told me. "My kids make fun of me all the time, and my husband rolls his eyes. It's because the littlest things seem to make me sick. When my clothes come back from the dry cleaner, I can smell the chemicals on them. Actually, I've given up buying clothes that need dry cleaning, but last week my husband had his suit cleaned before we went to a wedding together, and I almost got sick driving with him. It was like the fumes filled up the car!"

I asked Megan what else provoked a strong reaction, and she made a face. "How much time do you have?" she asked, before rattling off a long list that included air fresheners, dryer sheets, spray cleaners, perfumes, shampoos, soaps, paints, pesticides, car exhaust, new asphalt, herbicides ...

"No matter where I go, I can't escape the assault of chemicals," she lamented. "And the smallest exposure to yet another one triggers me. My poor husband can't wear aftershave, let alone cologne. I buy fragrance-free shampoos and soaps for all of us, even the kids—my oldest daughter still hasn't forgiven me. And if we go over to visit my mom or my aunt and they have a scented candle burning, watch out!"

Suddenly, Megan's eyes filled with tears.

"I'm a burden to them," she said bleakly. "Not only are they tired of me 'policing the smells,' they think I'm making things up. But I *know* there's something going on. I've seen the pattern long enough! I have rough, red, itchy skin—it's so gross, and I *know* it gets worse when I've gotten a big whiff of something. I get headaches. I feel faint. I'm constantly short of breath. And I don't know if this is related, but just when I think I'm finally going to be able to lose some weight, I don't."

Megan took on a determined look. "Clean, fresh air. That's all I ask for," she said. "Why do we need all these chemicals?!"

GST and GPX: Detox Partners

These two detox genes, GST and GPX, get dirty in similar ways, and you keep them clean in similar ways. Both are absolutely vital to clearing problematic compounds from your system! Since I couldn't discuss one of them without bringing in the other, this chapter treats the two of them together.

Your GST/GPX in Action

If you recall Keri, whom you met in chapter 1—my patient whose eyes were constantly watering and whose nose was always running—you've seen one example of what a dirty GST or dirty GPX can do to you. Megan, with her supersensitive sense of smell and her intense reactions to chemical fumes, is another example. A dirty GST or GPX—whether it was born dirty or is only acting that way—is no picnic! If you feel that this profile describes you, let me say it loud and clear: *You are not making things up!* You're struggling with a gene that was either born dirty or is acting dirty—and it's one of the genes that help your body manage glutathione, a powerful substance we've encountered in earlier chapters.

When you've got the right levels of glutathione, you walk into the world protected. Your immune system has a strong defender by its side, a biochemical that keeps toxins and industrial chemicals from triggering an immune reaction. In the right amounts, glutathione helps your body cope with a world that increasingly seems to be drowning in industrial chemicals.

Every single day you come in contact with industrial chemicals and heavy metals—in your indoor air, your outdoor air, your food, your water, and the vast majority of home and office products: furniture, photocopier toner, carpeting, mattress, cookware, cleaning products, personal-care products (including shampoos, lotions, cosmetics), and most especially plastics, plastics, plastics, plastics. Plastics line the cans in which food is stored; they coat the receipts you get at the cash register; they're used to store and cook your food; they hold your "pure" and "filtered" designer water. Plastics are nearly impossible to avoid, and every time you come in contact with them, you experience yet another toxic exposure.

My MTHFR personality means that I tend to go from zero to sixty pretty darn quick, so I try not to give in to my anger. But when I think about the overload of chemicals being dumped into our environment, I can't help it: I get mad. Fighting mad.

Look, genetic polymorphisms—SNPs—have been around a long, long time. They were adapted through natural selection, because they helped our ancestors cope with various aspects of our environment.

But think about it. How long have we had chemicals, processed foods, harmful medications, high-stress jobs, rush-hour traffic, and superbugs?

One hundred years? Maybe one hundred and fifty?

But SNPs have been around way longer—very likely, for most of the time that we humans have been on this planet.

So what changed? Why are people sicker? Our genes haven't changed much at all—so it must be our environment, lifestyle choices, and food.

Now, make no mistake: environmental chemicals hurt everyone. If you're exposed to industrial chemicals and heavy metals long enough or in great enough quantities, you greatly increase your risk of serious chronic disease.

But if you were born with SNPs in the GST or GPX gene, you're going to feel the effects of these chemicals sooner—long before you have the chance to get sick. That gives you a head start on cleaning up your environment and protecting your health. And if your born-clean GST/GPX is acting dirty, that's another powerful heads-up, nudging you into action. Because what look like minor symptoms today can turn into major symptoms tomorrow. And that's no fun at all.

GST/GPX: The Basics

Primary function of the GST gene

The GST gene makes the GST enzyme, whose primary job is to help your body transfer *glutathione*—your body's chief detox agent—to *xenobiotics* (harmful environmental compounds such as pesticides, herbicides, and heavy metals) that have infiltrated your body, enabling you to pee them out. If not eliminated, these chemicals damage your DNA, cell membranes and mitochondria, enzymes, and proteins.

Effects of a dirty GST

With a dirty GST, your body is unable to attach glutathione to xenobiotics—an especially big problem if you're facing a lot of chemical exposure.

Signs of a dirty GST

Common signs include hypersensitivity to chemicals (which can create such responses as congestion, runny nose, watery eyes, coughing, sneezing, fatigue, migraine, rashes, hives, digestive issues, anxiety, depression, and brain fog), increased inflammation, high blood pressure, and overweight/obesity.

Potential strengths of a dirty GST

Although everyone is vulnerable to industrial chemicals, your increased vulnerability makes you aware of the problem sooner and builds motivation for you to protect your health. You also have a better response to chemotherapy, since your GST can't easily clear these chemicals from your system.

Primary function of the GPX gene

The GPX gene makes the GPX enzyme, which helps attach glutathione to hydrogen peroxide (which is produced in your body as a by-product of the stress response), thus converting it to water that you can pee out.

Effects of a dirty GPX

With a dirty GPX, you can't efficiently use glutathione to convert hydrogen peroxide to water. Excess hydrogen peroxide disrupts your Methylation Cycle.

Signs of a dirty GPX

Common signs include early graying or white hair, erratic moods, chronic fatigue, memory problems, irritability, and aggression.

Potential strengths of a dirty GPX

Your increased vulnerability to excess hydrogen peroxide makes you aware sooner of the problem and more motivated to do something about it.

Going Gray: Part 2

As we saw in the previous chapter, your body makes hydrogen peroxide when your MAOA removes stress neurotransmitters from your body. The more stress you're under, the more hydrogen peroxide your body makes. That excess hydrogen peroxide can discolor and damage your hair.

Luckily, you have GPX to enable glutathione to convert hydrogen peroxide to harmless water. Lacking sufficient glutathione, you face all sorts of potential damage—not just to your hair, but particularly to your brain.

For that reason, you don't want to overburden your GPX and cause it to get dirty. This is why stress relief is so important: the more stressed you are, the more hydrogen peroxide you release and the more glutathione you need.

It's also why diet is so important. Under stress, most of us tend to binge on carbs, high-fat foods, and sugars, all of which deplete glutathione even more.

How about that for a vicious cycle? Stress increases hydrogen peroxide, and then you crave carbs which *further* increase it. Stress and carbs deplete your glutathione dangerously. Unfortunately, they're not the only threats.

The problem becomes even more pressing when you have an infection of any type—viral, bacterial, mold, yeast, or parasite. At those times, your immune system is fighting hard for you, and one of its weapons in fighting off infections is hydrogen peroxide. This means that any time you get sick or have a chronic infection, you burn through your glutathione leaving you and your body susceptible to damage.

Meet Your Dirty GST/GPX

If you've been through Laundry List 1, you already have a good idea about whether your GST or GPX gene is dirty. But here are a few more questions to help you figure that out:

☐ I am (or have been) infertile.
☐ I'm sensitive to chemicals and smells.
☐ I feel better after a sauna or intensive exercise.
☐ It's easy for me to gain weight even though I eat right.
☐ Cancer runs in my family.

Yeah. Cancer. I don't want to scare you. In fact, I want to get you as jazzed as I am about how much you can do to clean up a dirty GST/GPX. But if you don't, cancer is one potential outcome. Let's do everything we can to keep that from happening.

Health Conditions Related to a Dirty GST/GPX

The list of health conditions that researchers have linked to a dirty GST/GPX is daunting.

- Alzheimer's disease
- Amyotrophic lateral sclerosis (ALS)
- Anxiety
- Autism
- Autoimmune conditions, including Graves' disease, Hashimoto's thyroiditis, multiple sclerosis, rheumatoid arthritis

- Cancer
- Chemical sensitivity
- Chronic infections such as hepatitis, mold reaction, Epstein-Barr, *Helicobacter pylori*, and Lyme disease
- Crohn's disease
- Depression
- Diabetes, types 1 and 2
- Eczema
- Fatigue
- Fibromyalgia
- Heart disease
- Hypertension
- Hearing loss
- Homocysteine surplus
- Infertility
- Keshan disease (a type of heart problem)
- Mental disorders, including major depressive disorder, bipolar disorder, schizophrenia, and obsessive-compulsive disorder
- Migraine
- Obesity
- Parkinson's disease
- Pregnancy complications
- Psoriasis
- Seizure
- Stroke
- Ulcerative colitis
- Vision loss (progressive worsening)

GST and Your Microbiome

There are many types of GST gene, each with its own unique job. They reside mainly in the intestines and liver—but your microbiome also has its own GST enzymes. In fact, your microbiome is a key player in your body's effort to get rid of xenobiotics, protecting you against chemical and oxidative stress. Think of your microbiome as your GST's main backup—and make sure to protect it!

What Makes GST/GPX Dirty?

- **Exposure to a lot of industrial chemicals, heavy metals, bacterial toxins, and plastics.** The more you lighten the chemical burden on your GST/GPX, the better chance you give this gene to function at its best.
- **Stress.** Under physical and mental stress, your Methylation Cycle uses up more ingredients than intended, yet doesn't function as well as it should, which means your body is short of the raw materials it needs to make glutathione. Stress dirties all your genes surprisingly quickly—including your GST/GPX.
- **A disrupted Methylation Cycle.** Whenever your Methylation Cycle struggles, you have trouble making all the glutathione your body needs. This puts a big strain on your GST/GPX.
- **Insufficient riboflavin/vitamin B$_2$.** Your body uses riboflavin to regenerate decaying, dysfunctional glutathione back into whole, functional glutathione. If you're not consuming enough foods that are rich in riboflavin, your supply of glutathione can't keep up. Without functional, healthy glutathione, you can't remove industrial chemicals or hydrogen peroxide from your body. And your GST/GPX has to work harder to combat the chemical onslaught.
- **Insufficient selenium.** In order for your glutathione to turn hydrogen peroxide into water, it needs selenium. Without selenium, your GPX enzyme can't get rid of hydrogen peroxide.
- **Insufficient cysteine.** Cysteine, found in many nutritious foods and made from your homocysteine, is the key

ingredient in glutathione. And as you know by now, if your GST/GPX gene doesn't have enough glutathione available, it can't function at all.

Glutathione as Primary Protector

You've probably heard a lot about the benefits of antioxidants, and perhaps you noticed that I referred to glutathione in an earlier chapter as an antioxidant. In fact, I'm here to tell you that glutathione is your *primary* antioxidant. But what does *antioxidant* mean, and why is oxidation a bad thing?

Good question. First, a one-sentence answer. Our bodies burn oxygen for fuel, which is a good thing—but that process produces various harmful chemicals. (One of those harmful chemicals is hydrogen peroxide, which, as we've seen, is rendered harmless by glutathione.)

Now for a more detailed answer. Your body burns oxygen within your *mitochondria,* which are the energy powerhouses of your body's cells. From this process, your mitochondria produce your body's primary energy carrier, *adenosine triphosphate* (ATP). But the burning process produces lots of harmful by-products, including free radicals. To protect themselves against those by-products—the source of what we call *oxidative stress*—your mitochondria need lots of glutathione. Otherwise, they become damaged and can't produce enough ATP. When that happens, your cells don't get the energy you need, and you end up with a myriad of conditions (see the list above). Long story short: glutathione is crucial to your overall function.

Guess what else produces damage for glutathione to clean up? Inflammatory foods—especially sugars and unhealthy fats—and just plain eating too much—of anything. Overeating produces an inflammatory compound called *methylglyoxal*—which is commonly elevated in diabetics, people on a high-protein diet, and people following a ketogenic diet. Glutathione protects you by converting methylglyoxal into harmless lactic acid.

Glutathione and Your Weight

The more industrial chemicals, oxidative stress, and toxins in your system, the more excess weight you're likely to carry. For that reason, cleansing your body of its toxic burden is likely to make you slimmer. I've had clients who lost five or ten pounds just by reducing their chemical exposure. The biochemistry is intricate, but the conclusion is simple: when you're not overburdening your genes or depleting your stores of glutathione, you'll find it much easier to achieve your optimal weight.

How is this possible? Think of food the way your cells and genes do—not as a good taste or a filling sensation, but as fuel and tool. You achieve an optimal weight when your mitochondria are able to burn the calories you ingest as fuel. If you have low glutathione levels, your mitochondria can't work very well. So where does that unburned fuel go? Your waist. Give your mitochondria the glutathione they need—and you maintain an optimal weight. It's a lifelong partnership.

Glutathione and Vitamin B$_{12}$

Vitamin B$_{12}$ is essential for preventing anemia, supporting your cells with oxygen, and preventing nerve damage.

But it's not enough just to *consume* B$_{12}$. You need carrier proteins to transport it into your cells—and glutathione is the glue that helps B$_{12}$ stick to the carriers. So if you're low on glutathione, you can take all the B$_{12}$ supplements you want and it won't be transported to the places where your body needs it—thus your B$_{12}$ deficiency will continue. Once again, glutathione is the key.

Glutathione and Your Methylation Cycle

Your Methylation Cycle depends on glutathione, as we saw in chapter 5. The moment hydrogen peroxide levels rise and heavy metals accumulate, your Methylation Cycle stops. If your GST/GPX gene is dirty, then your Methylation Cycle is dirty. Glutathione is the key to healthy methylation.

Glutathione and Your Brain

In order for your brain to make dopamine and serotonin, you need glutathione on board. The moment your glutathione levels drop, your ability to make these vital neurotransmitters does, too. No wonder low glutathione levels are associated with so many mental health and neurological conditions, including ALS, major depressive disorder, bipolar disorder, drug addiction, obsessive-compulsive disorder, autism, schizophrenia, and Alzheimer's disease.

Glutathione and Your Heart

An essential compound for a healthy heart and blood vessels is nitric oxide. When glutathione levels drop, your ability to make nitric oxide goes down; and, as a result, your heart and blood vessels don't function as well as they need to. So glutathione is essential for your heart health.

Glutathione and Your Immune System

Glutathione helps your immune system fight infections effectively. When glutathione levels drop, people typically experience autoimmunity—that is, their body fights not infection, but itself. The result is inflammation. In short, then, without glutathione, your immune system is ineffective at eliminating infections, and your inflammation is high.

Furthermore, your immune system, like your heart, needs nitric oxide; it uses that compound to fight infections. When glutathione is low, your body's ability to generate nitric oxide for fighting infections drops, allowing infections to persist.

Key Nutrients for a Healthy GST/GPX

As we've seen, the GST/GPX gene's work involves transferring the antioxidant glutathione to chemicals and compounds that need to be eliminated from the body. In order to make that antioxidant, your body requires *cysteine*, a sulfur-containing amino acid that many people are deficient in:

Cysteine: red meat, sunflower seeds, chicken, turkey, eggs, broccoli, cabbage, cauliflower, asparagus, artichoke, onions

You also need riboflavin to transform damaged glutathione back into a ready-to-use antioxidant. Otherwise, damaged glutathione remains damaged—and contributes to further damage in your cells.

Riboflavin/B$_2$: liver, lamb, mushrooms, spinach, almonds, wild
    salmon, eggs

Finally, your GPX needs *selenium,* a trace mineral that many people
are deficient in:

Selenium: brazil nuts, tuna, halibut, sardines, beef, liver, chicken,
    brown rice, eggs

Balancing Your Sulfur

Your body needs a *lot* of sulfur, which it uses for blood flow, healthy joints,
gut lining repair, and the elimination of hormones and neurotransmitters. You also need sulfur to make glutathione. That sulfur comes primarily from dietary protein and cruciferous vegetables, including foods
rich in cysteine (the sulfur-containing amino acid mentioned above).

Unfortunately, some people can't seem to tolerate sulfur well. An
unfriendly microbiome may be the culprit—one that's producing too
much hydrogen sulfide. If you smell like rotten eggs, with a sulfur odor
coming from your armpits, breath, stool, and gas, then you're probably
high in hydrogen sulfide. (Loose stool is another common, but not inev-
itable, sign of this condition.)

If this is the case, cut back on high-sulfur cruciferous vegetables
(broccoli, brussels sprouts, cabbage, cauliflower, and kale), and cut back
on any sulfur-based supplements you're taking, such as MSM (methyl-
sulfonylmethane) or NAC (N-acetyl cysteine). Ask your doctor to order
a comprehensive digestive stool analysis (CDSA) to see what's happening with your microbiome.

Whatever the cause of sulfur intolerance, eliminating sulfur from the
diet isn't a cure-all. People on a low-sulfur diet may feel better initially—
but in the long term may end up with a significant sulfur deficiency.

I've had countless clients who had difficulty balancing the sulfur in
their diet. For example, Janet had chronic headaches, dizziness, pain
everywhere, and nosebleeds, plus she felt horrible after eating. I reviewed

her diet and supplements and noticed that she was ingesting a lot of sulfur. She happened to be following the high-protein GAPS diet. On top of that, she was taking an MSM supplement for her joint pain and gut repair and an NAC supplement to help boost her low glutathione levels, both of which drove her sulfur levels even higher. She was overdosing on sulfur!

Stopping the sulfur-containing supplements and reducing her protein intake lowered her sulfur levels and eliminated her symptoms. But two months later, Janet was back.

"I feel like I'm starving for air," she told me. "I can't breathe! I'm devastated—and it keeps getting worse."

I told her that now her sulfur levels were *too* low. In order to be able to breathe, her lungs need adequate hydrogen sulfide—not too much, but not too little.

"You're experiencing a yo-yo effect," I explained. "You're putting your body at extremes. And now, the longer you stay on a low-sulfur diet, the more your glutathione supply will become depleted. That's because your body will break apart glutathione to pull out the additional sulfur that it needs. So a low-sulfur diet means you end up short of sulfur *and* glutathione."

I figured out a new dietary plan for Janet and taught her the Pulse Method of supplementation—which you will learn in chapter 12. The Pulse Method enables you to figure out when you need to keep taking a supplement—and when you're ready to change the dose or stop altogether.

Two days later, I heard from Janet: "Magic. Just magic. I can breathe! I understand the Pulse Method now. I'll keep an eye on how I'm feeling, and I'll tune in and adjust my diet and supplements as needed. I feel so empowered. Thank you! Thank you!"

Sulfite Sensitivity

Sulfites are sulfur compounds that occur in some foods naturally and are added to others for their antioxidant and preservative properties. (Sulfites are often added to wine, for example, and to dried fruits.) Just as some people can't tolerate sulfur, many have a sulfite sensitivity; for them, sulfites are allergens and must be avoided.

Whether a person has a sulfite sensitivity or not, sulfites must be eliminated from the body. If they're allowed to build up, health problems, including asthma, can result.

Enter SUOX, another key gene. It's not one that we're focusing on in this book, but because it's a partner in the detox business, we'll touch on it here. SUOX uses molybdenum, a dietary mineral, to eliminate sulfites from the body.

A high-protein diet and/or high-sulfur supplement strains that gene. That's exactly what happened to Janet during her high-sulfur phase: she overwhelmed her SUOX gene and her body ran out of molybdenum, causing her sulfite levels to rise.

Making the Most of GST/GPX

Just as with Keri, I wanted Megan to know that there was hope for her—plenty of hope. She was in no way doomed to a life of skin rashes and headaches and mockery from her family.

I reminded Megan that the more xenobiotics, free radicals, reactive oxygen species, sugar, excess fat, and excess protein she was exposed to, the more glutathione she would need. Producing and recycling glutathione is a demanding and difficult process that requires a number of genes and enzymes. So cleaning up her environment and her diet would make a terrific start to easing her symptoms and cleaning up her genes.

I also reminded Megan that the dirtier her glutathione genes—GST and GPX—the less well her cells were going to work. And it was poor cell function that had created her chronic symptoms.

Here are some of the ways that Megan could begin cleaning up her genes—and you can, too. You don't even have to wait for your Soak and Scrub. Just jump right in:

- **Eat a lot of fiber.** Your microbiome loves fiber! Those gut bacteria eat the fiber that your own body can't digest, and then they help your body detox. Fiber contributes to the production of detoxification enzymes, and it also binds to xenobiotics. Once fiber hooks up with those chemicals, it ushers them out through your stool. Problem solved! Exception: If you're struggling with SIBO, you should *not* start eating more fiber. You'll have to address the SIBO first.

High-Fiber Foods

- Artichokes
- Avocados
- Black beans
- Blackberries
- Broccoli
- Brussels sprouts
- Chia seeds (which you can sprinkle on salads and vegetables or stir into yogurt)
- Flaxseed meal (which you can add to oats, smoothies, yogurt, and baked goods)
- Lentils
- Lima beans
- Oatmeal (stick to gluten-free)
- Pears
- Peas
- Raspberries
- Split peas

- **Clean up your environment.** Every time you eat, drink, breathe, or touch an industrial chemical—and yes, I'm including plastics, pesticides, air fresheners, dryer sheets, herbicides, and car exhaust—you add another load to your body's burden. The more you can limit your exposure, the less detox labor your body has to do, the less glutathione you need, and the easier job your GST/GPX has. You don't want to overwork any GST/GPX, but especially not one that was born dirty and already has trouble getting the job done. Remain on the Clean Genes Protocol as you filter your water, eat organic food, clean your indoor air (especially at home), and avoid toxic products.

- **Evaluate your environment for mold.** If you're struggling with lots of symptoms that won't go away even after you clean up your diet, air, water, and products, you might want to test for mold in your home, at work, in your car, and anywhere else you spend a substantial amount of time. An environmental inspector can help make this evaluation.

- **Sweat it out.** Your body detoxes in four ways—breath, pee, poop, and sweat. You're already breathing—and hopefully properly—so that takes care of that. You're already hydrating, so that takes care of the pee. You're eating lots of fiber—that takes care of the poop. Now let's get you breaking a sweat at least two times a week. You've got lots of choices, from the energizing to the super-relaxing: sauna, Epsom salt bath, vigorous exercise, hot yoga, sex. If you go the sauna route, choose low heat so that you can stay in there longer and keep sweating. Living in a hot climate like Arizona is not going to cut it unless you get outside and sweat. Because with a GST/GPX SNP, you've gotta sweat. A lot.

- **Be aware.** Know that you're sensitive to chemicals, and keep avoiding them. At the same time, know that others aren't as sensitive as you, which may be why they don't follow your recommendations or believe your stories about how chemicals are affecting you. It can be tricky to convince a skeptical family or a doubting friend—but the first step is for *you* to believe yourself.

- **Grow broccoli sprouts and radish sprouts.** I'll warn you—the taste is quite potent! But you'll get tremendous glutathione support. It's the combination of sprout types that does it. Eating broccoli sprouts on the third day after they sprout gives maximum benefits.

Just as Keri found a lot of success cleaning up her GST, so did Megan enjoy improved health from cleaning her GST/GPX. She had already done a solid job of reducing her exposure to toxic chemicals, but with my help she identified a few more that she had missed.

What she hadn't done was support her body's detox through breath, pee, poop, and sweat. So she began focusing on breathing properly, hydrating regularly, eating more fiber, and taking twice-weekly saunas. "I can *feel* the toxins flowing out of my body," she told me at our second meeting. "It's the best thing—and also so relaxing!"

Megan also found a lot of support in living according to the Clean Genes Protocol: especially getting better sleep, and reducing and relieving stress. These key steps helped support her entire system, lightening the burden on all her genes and allowing her dirty GST/GPX to function at its highest capacity.

Remember when I told you that a lot of Italians have a dirty MTHFR but live symptom-free, without supplements or medications? That's possible for people with a dirty GST/GPX also—as long as you're eating well, getting the right kind of exercise, enjoying deep sleep, avoiding toxins wherever possible, and reducing or relieving stress. This is the protocol I follow, and my family too; and we've seen it ease the burden on our dirty genes. Keri and Megan found success with it as well—and I know you can too.

How the Clean Genes Protocol Supports Your GST/GPX

Diet. You'll get a balance of fiber, sulfur, and riboflavin/vitamin B_2, which will support your body's stores of glutathione as well as the function of your GST and GPX. You'll eat healthy fats and cut out the processed carbs, sugars, and unhealthy fats that trigger a number of glutathione issues and burden your GPX. And you'll eat the right amount of protein, rather than burdening your body—and your genes—with too much of it.

Chemicals. Avoiding industrial chemicals will lighten the burden on both genes and on your stores of glutathione. You'll be breathing, sweating, peeing, and pooping out the toxins, which will help as well.

Stress. Stress relief lightens the burden on your MAOA, which in turn decreases the amount of hydrogen peroxide your body makes, thereby lightening the burden on your GPX. Identifying infections and eliminating them also significantly reduces the amount of hydrogen peroxide generated and reduces the work your GPX has to do.

10

NOS3: Heart Issues

Rudy was a big man who used to work construction. He took early retirement because of some injuries and noticed that as soon as he wasn't as physically active as he used to be, his blood pressure started to creep up. Although he had suffered the occasional migraine over the years, now he was getting one once a week.

Rudy had a dirty NOS3, a gene frequently associated with cardiovascular issues and migraines. Like Jamal, whom you met in chapter 1, Rudy was especially concerned because heart disease ran in his family—his grandfather had died of a heart attack, his father had struggled with high blood pressure, and one of his uncles had died of a stroke.

I reassured Rudy that we could reverse this apparent "genetic destiny" by cleaning up his genes, especially NOS3. NOS3 is a gene central to the making of nitric oxide, a substance that keeps our blood vessels dilated. When NOS3 gets dirty, though, it doesn't make nitric oxide as efficiently as it should. As a result, your blood vessels become constricted and can't properly deliver the oxygen that's supposed to travel through your blood.

"Think of it this way," I told Rudy. "Your cells need to breathe just like you do, or too many of them die. Your bloodstream carries blood—and oxygen—to all of your cells. So if your blood vessels are constricted, your cells don't get enough blood—or oxygen—and then they can't breathe."

Rudy nodded.

"Okay," I continued. "Your heart uses the most oxygen of any part of your body, ounce for ounce—even when you're sitting quietly at rest. So what happens if your heart cells can't breathe? Without the oxygen they need, a lot of cells die—and if enough die, you might get angina, or maybe even have a heart attack."

I went on to explain that the next biggest user of oxygen by mass is your brain. If your brain cells don't get enough oxygen, they can't breathe either. If too many brain cells die, you might develop a migraine or, in severe cases, maybe even brain damage.

"Bottom line," I told Rudy, "right now your blood vessels aren't properly dilated, so they can't deliver enough blood—which means they can't deliver enough oxygen either. That's the situation that we have to turn around."

Rudy was following me closely, but I wasn't done yet.

"Now, another thing that happens with low nitric oxide is that your blood platelets—the components that help blood clot in the case of emergency—get 'sticky.' As your blood travels through your arteries, you want it to flow smoothly. If the platelets in your blood start sticking together, they can form a clot when no clot is needed. This typically is a slow, stealthy process—but that's how your uncle ended up having a stroke."

Rudy was still with me, but he looked shaken.

"I'm telling you what *might* happen if we *don't* clean up your genes," I reminded him. "Don't worry—we're *going* to clean up your NOS3. But there's one more thing I want you to understand. A dirty NOS3 means that you're slow to produce new blood vessels. The scientific term for that formation is *angiogenesis*. If you don't have healthy angiogenesis and you're wounded—say, from a cut or a deep scratch—your body struggles to make the extra blood vessels that carry the nutrients and oxygen you need to repair the wound. So your wound will heal a lot more slowly."

Rudy nodded again. "I did get some deep cuts and scratches on the job," he said, "and the doctor told me they were taking longer than usual to heal."

"Makes sense," I said. "But remember, we can turn this all around."

Now Rudy had a question. "What about my high blood pressure? How is that related?"

"When your blood vessels don't dilate properly, the blood flowing through them puts more pressure on the vessel walls," I explained. "This can happen even when you're otherwise in pretty good health. It's called *essential hypertension*. Before you retired, when you were more active, you were breathing in more oxygen, and that was helping to keep your blood vessels dilated, even with your dirty NOS3. Now that you're less active, your dirty NOS3 is really showing its effects."

As Rudy was learning, dirty NOS3 is a terrific example of why it's important to know your dirty genes—and why it's even more important to know how to support them. Yes, if you've got a genetic propensity for cardiovascular disease, it might be life-threatening—but it doesn't have to be. The right diet, supplements, and lifestyle can make all the difference.

Your NOS3 in Action

As we've seen, the NOS3 gene works hard to keep your heart and your vast circulatory system healthy, which impacts all the organs served by that circulatory system. Pretty crucial stuff.

Interestingly, depression is an independent risk factor that doctors use to evaluate whether you're at risk for cardiovascular disease. That's because depression is often associated with low levels of dopamine and serotonin. Dopamine, you may recall, is the neurotransmitter that revs you up, gets you ready to meet challenges, and enables you to enjoy such thrills as a roller-coaster ride or falling in love. Serotonin is the neurotransmitter that supports your optimism, calm, and self-confidence.

Chemicals are a primary way that your NOS3 gene gets dirty. Although effects of chemicals on your circulatory system can't be sensed in the moment, effects on mood differences can. Next time you're exposed to chemicals of any type, see if the exposure affects your mood. If so, you're witnessing NOS3 interacting with your brain chemistry. We will explore why this happens later in this chapter.

NOS3: The Basics

Primary function of the NOS3 gene

The NOS3 gene influences the production of nitric oxide, which is a major factor in heart health, affecting such processes as blood flow and blood vessel formation.

Effects of a dirty NOS3

With a dirty NOS3, you don't produce enough nitric oxide. As a result, your blood vessels don't dilate sufficiently and your platelets can become sticky, which can lead to blood clots.

Signs of a dirty NOS3

Common signs include angina, anxiety, cold hands and feet, depression, heart attack, erectile dysfunction, high blood pressure, migraines, mouth-breathing, sinus congestion, and wounds that are slow to heal.

Potential strengths of a dirty NOS3

Potential strengths include decreased blood vessel formation during cancer, which reduces the growth of cancer.

Meet Your Dirty NOS3

You've just seen that a dirty NOS3 can lead to high blood pressure, cardiovascular issues, blood clots, and stroke, as well as depression. It can also produce complications for diabetics.

Diabetes is known to cause significant difficulties in blood flow and healing. Legs are cold. Ulcers form. Toes have to be amputated. Diabetes is also known to cause loss of vision. All of these issues result from a dirty NOS3: insufficient nitric oxide leads to a loss of blood flow; as a result, your legs, feet, and eyes can't get the nutrients and oxygen they need.

Why not? Well, when you have diabetes, your blood levels of insulin are high all the time. And, among other things, insulin pushes NOS3 to make nitric oxide.

That's usually a *good* thing, and in healthy people it remains so. But diabetes dirties up your NOS3 if it wasn't already born dirty. So instead of making nitric oxide, your NOS3 makes *superoxide,* one of the most dangerous free radicals there is. This reactive compound causes all kinds of havoc in your body—and diabetic complications are the result.

Another NOS3 danger is birth defects. During fetal development, your baby is growing rapidly and needs you to form new blood vessels to nourish his or her developing cells and tissues. If a dirty NOS3 slows down your ability to form these blood vessels, your baby's heart won't get the support it needs and he or she could develop a congenital heart defect—which, as it happens, is the most common birth defect in humans.

So *yes,* good thing you're going to learn how to clean up your dirty NOS3, if you have one! Your Laundry List 1 has given you some indication, but here are a few more factors to help you determine whether your NOS3 is either born dirty or acting dirty:

☐ I have high blood pressure.

☐ Many people in my family have high blood pressure.

☐ Heart attacks are common in my family.

☐ I've had a heart attack.

☐ I have a lot of circulation issues due to my diabetes.

☐ I frequently have cold hands and feet.

☐ Stroke runs in my family.

☐ I was diagnosed with preeclampsia when I was pregnant.

☐ Hardening of the arteries (atherosclerosis) runs in my family.

☐ I'm a mouth-breather.

Health Conditions Related to a Dirty NOS3

Whether your NOS3 was born dirty or is only acting dirty, it sets you up for some potentially serious disorders, contributing to more than four hundred conditions. Here are the ones most closely associated with a dirty NOS3.

- Alzheimer's disease
- Angina
- Asthma
- Atherosclerosis
- Bipolar disorder
- Brain ischemia
- Breast cancer
- Cardiovascular disease
- Carotid artery disease
- Chronic sinus congestion
- Coronary artery disease
- Depression
- Diabetes, types 1 and 2
- Diabetic nephropathy
- Diabetic retinopathy
- Erectile dysfunction (often an early sign of cardiovascular disease)
- Hypertension
- Hypertrophy, left ventricular
- Inflammation
- Kidney failure, chronic
- Metabolic syndrome (or syndrome X)
- Miscarriage, recurrent
- Myocardial infarction
- Neurological disorders, including ALS
- Obesity
- Preeclampsia
- Prostate cancer
- Pulmonary hypertension
- Schizophrenia
- Sleep apnea
- Snoring
- Stroke

NOS3 Linkages

Let's take a closer look at some of the causes and effects of a dirty NOS3.

Sinus Congestion and Runny Nose

A congested or runny nose has been identified as a possible contributor to high blood pressure. That's because if you aren't breathing in enough oxygen, you're making your NOS3 dirty.

Having a congested nose doesn't mean you should run out and get nasal spray. It means you need to identify the source of the problem and remove it. Perhaps a dirty DAO? Or a reaction to dairy products? Another type of food or environmental sensitivity?

Or perhaps the culprit is a dirty NOS3. Sinus congestion can be due to low levels of nitric oxide. We don't want this minor breathing problem turning into high blood pressure.

Cold Hands and Feet

Many of us struggle with cold hands and feet. Oh, what we'd give to warm them up naturally so we didn't have to always wear gloves or hear someone say, "Oh my—your hands are freezing!"

Cold extremities are a sign that your NOS3 is dirty. If your blood can't get out to your fingers and toes, it means that your blood vessels are too constricted. Cleaning your dirty NOS3 should make a big difference.

Mouth-Breathing

Mouth-breathing is a very ineffective way to oxygenate your body. And low oxygen levels lead to a very dirty NOS3.

Mouth-breathing happens for a number of reasons. One possibility is that sinus congestion is forcing you to breathe through your mouth. In that case, addressing your congestion (see above) should resolve the problem.

Assessing your home for mold, evaluating food allergies or intolerances, and fixing your dirty DAO are other great ways to address sinus congestion. Cleaning your dirty NOS3 will also resolve some types of congestion.

Yet another possibility is that your sinuses are blocked by nasal polyps, especially if you feel that air flow into your nose is uneven. Nasal polyps are commonly associated with persistent allergies—either environmental or food. They can be surgically removed, but they might come right back if the allergies aren't addressed.

A deviated septum is another common reason for mouth-breathing. Your doctor likely has informed you of this problem, if your septum is affected. What he or she probably hasn't told you is that you should have it fixed so that you can breathe properly. NeuroCranial Restructuring, a technique that involves adjusting the cranial plates through the sinuses, is an effective nonsurgical way to fix most deviated septum types.

Other causes for mouth-breathing are related to facial structure. A condition known as tongue-tie is incredibly common. An anomaly in how the tongue is attached to the bottom of the mouth, it alters facial structure, which leads to mouth-breathing. If your baby or young child is mouth-breathing, have him or her evaluated by a lactation consultant. There are various types of tongue-ties—anterior tongue-ties (which are easy to spot), posterior tongue-ties (which are tougher to identify), and upper- or lower-lip ties (which are fairly visible). If your child has an improper latch during breastfeeding, difficulty saying some words, or trouble swallowing food or pills, tongue-tie may be to blame.

It's possible to correct a tongue-tie—ideally at birth but even as an adult—so check with your dentist. As was mentioned, many lactation consultants are also very knowledgeable about ties. Tongue-tie can sometimes be corrected by a simple snip, but more often laser treatment is needed. The results are phenomenal: improved breathing, easier breastfeeding, more fluid speech, easier swallowing—and a happier NOS3 gene.

Pollution, Smoking, and Stress

Stress, smoking, and pollution can dirty up your NOS3 even if it was born clean. That's because NOS3 relies upon a compound known as BH4, which your body produces. BH4 loves things clean. If your body is stressed by or dirtied up with toxins—including nicotine and industrial chemicals—your BH4 levels will go way down.

Without BH4, though, your NOS3 can't generate nitric oxide. Instead, as we saw earlier, your NOS3 produces superoxide—that dangerous free radical associated with diabetic complications. Unfortunately, it's not only diabetics who have to worry about diminished BH4. If you're lacking in BH4, you'll also end up with reduced blood flow, stickier blood platelets, and a greater risk of cardiovascular disorders—whether you're diabetic or not.

NOS3 and Neurological Disorders

As mood disorders persist in an individual, they tend to become deeper, more entrenched. Eventually, they may even result in neurological disorders such as Parkinson's, ALS, or seizures. Just as depression is linked to cardiovascular disorders, so is it linked to disorders of the nervous system. If BH4 continues to be in short supply and leads to the production of superoxide, your brain, the "director" of the nervous system, becomes persistently—and insidiously—harmed. Please tune in and catch these signs early!

NOS3 Impacts for Women

A dirty NOS3 is of particular concern for pregnant women and for post-menopausal women. Let's look at why.

NOS3 in Pregnant Women

During pregnancy, women experience high levels of estrogen and nitric oxide. In fact, estrogen stimulates NOS3 to work better and to produce more nitric oxide. This additional nitric oxide is essential for forming new blood vessels, preventing blood clots, and increasing blood flow to the developing baby.

If you have a dirty NOS3 during pregnancy, you're at increased risk of recurrent miscarriage, congenital birth defects, and preeclampsia. I want you to know about these risks ahead of time so that you can support your NOS3 as needed and prepare for a safe pregnancy.

NOS3 in Postmenopausal Women

The risk of all types of heart disease—high blood pressure, blood clotting (stroke), and heart attack—increases dramatically for women after menopause. That's because estrogen, as noted above, stimulates NOS3 to produce nitric oxide. When estrogen levels drop after menopause, nitric oxide production falls and cardiovascular risk increases. This is yet another incentive for keeping your estrogen balanced and at a healthy level.

Stay Away from Statins

The class of drugs known as *statins* help stimulate the production of nitric oxide and support NOS3. They are among the most prescribed drugs in the United States, used by many doctors to lower cholesterol. But I'm always skeptical of relying on a drug to do something that your body is supposed to do by itself. After all, none of us is born with a statin deficiency.

Furthermore, a number of serious side effects have been associated with statins, including:

- Abdominal cramping or pain
- Bloating
- Constipation
- Diarrhea
- Dizziness
- Drowsiness
- Gas
- Headache
- Muscle ache, weakness, or tenderness
- Nausea or vomiting
- Rash
- Skin flushing
- Sleep issues

Statins can also produce even scarier side effects, especially among the elderly, including memory issues, mental confusion, increased blood sugar, and type 2 diabetes.

Given everything, wouldn't you rather look for natural ways to accomplish what statins are supposed to do? Especially since, as research suggests, statins don't seem to work well if your NOS3 is dirty.

The Nitroglycerin Connection

Your NOS3 is intended to give you all the nitric oxide you need. But when it doesn't function optimally, your doctor might prescribe nitro-glycerine. Yes, that's the same nitroglycerine used to blow things up in movies! Used short-term, as a quick fix, nitroglycerine can save someone's life. I don't like it as a long-term solution for heart issues, however, so let's take a closer look.

Nitroglycerin promotes the release of nitric oxide—*the* compound to support blood flow. That's terrific: now your cells are getting the oxygen and nutrients they need.

But sometimes nitroglycerin fails to work. Some people never respond to it; others develop nitroglycerin resistance.

Why the variance? I bet you've already guessed it—because of a dirty NOS3. If your NOS3 is only a little dirty and needs just a little support, nitroglycerin can help. If your NOS3 is very dirty and needs a lot of support, even a big blast of nitro won't be enough to get sufficient quantities of nitric oxide flowing. That's why smokers don't typically have success with nitroglycerin.

So I'm all for nitroglycerin as a short-term solution; it's literally a lifesaver. Long-term, though, you have to clean up your NOS3 along with the rest of your genes.

Meanwhile, if you're taking nitroglycerin and you start noticing that it isn't helping you as much as it used to, you need to inform your health professional—let him or her know that your NOS3 enzyme may be "uncoupling," a term I'll explain in the next section.

The Arginine Steal

Just as many doctors rely on nitroglycerine to treat heart issues, so do many others rely on a compound known as *arginine,* a type of amino acid found in both animal and vegetable proteins. Arginine *does* support a clean NOS3—but, like nitroglycerin, it doesn't necessarily work if you've got a dirty NOS3. In fact, both nitroglycerin and arginine can make your heart worse if your NOS3 is uncoupled.

An *uncoupled* NOS3 is one that's operating with insufficient arginine and BH4. Instead of making nitric oxide, which your blood vessels can use, an uncoupled NOS3 makes superoxide, which we've seen is highly dangerous. Industrial chemicals damage BH4, but how do you end up with a shortage of arginine?

Well, your body needs arginine for many purposes—not just to support your NOS3. For example, when your body is fighting an infection and is inflamed, genes directly involved in that fight need more arginine than usual. Those needy genes get it by "stealing" it from other genes, including NOS3. As less and less arginine is available to your NOS3, it stops making nitric oxide and makes superoxide instead. Superoxide damages BH4, so now you're low in BH4, too. And since arginine and BH4 are *both* low, you've got even more superoxide being made by your dirty NOS3. A dirty NOS3 just got dirtier.

As if that weren't enough, certain types of bacteria in your microbiome also use a significant amount of arginine, further "stealing" it from your NOS3. Yet another reason to evaluate your microbiome.

Now, at this point, you might be thinking, "Okay, fine. I'll just take an arginine supplement."

Can you guess why that doesn't work? I bet you know by now!

NOS3 needs *both* arginine *and* BH4. And remember, BH4 is supersensitive. It behaves the way you might if you found a bug in the food you ordered. Stops you 100 percent from eating it, just as the slightest bit of grime in your genes causes your BH4 to grind to a halt. So if you take arginine while your BH4 levels go down, all you're going to produce is superoxide, which is definitely going to make things worse. In fact, researchers have tried giving arginine to individuals with high blood pressure in order to increase nitric oxide. It didn't work.

Researchers have also tried giving BH4 to see if they could support NOS3 and nitric oxide production. That supplementation helped some people, but others found no benefit.

Here's how I look at it. If a building is on fire, you don't put in some new furniture—it's only going to get burned up along with everything else. By the same token, there's no point in taking supplemental BH4 if your blood and system are dirty—you'll just dirty up the supplemental BH4.

What *should* you do to support NOS3? There are three things you need to do—but you have to do all three, or none of them will work:

1. Supply adequate arginine.

2. Maintain a steady supply of clean BH4.

3. Keep all your other genes clean.

Now, you know I don't recommend starting out with an arginine supplement. But what if you've been taking arginine already? Perhaps it enhanced your performance in sports, reduced your headaches, and warmed your cold hands and feet, but now you're not noticing any benefit or have even found that you're worse. In that case, you might have an uncoupled NOS3. Stop taking the supplement immediately and work on cleaning up your NOS3.

How Your Body Uses Arginine

Arginine supports the following essential functions:
- Blood vessel dilation
- Creatine formation
- Infection fighting
- Immune tolerance
- Neurotransmission
- Penile erection
- Reduction in platelet stickiness

Folic Acid: Your Foe

You've already seen how bad folic acid is for your MTHFR and your Methylation Cycle. Well, it's also bad for your NOS3.

First, your NOS3 depends on a compound called NADPH, which folic acid also uses. So the more folic acid you consume, the more NADPH you pull away from supporting your NOS3. Second, as your level of folic acid increases, your level of BH4 decreases.

Remember, folic acid is synthetic. Our body was not designed to process folic acid. We *do* process it—but at great cost.

What Makes NOS3 Dirty?

- Breathing abnormalities
- Folic acid
- High blood sugar

- High carbohydrate intake
- High homocysteine levels
- High insulin levels
- Infection
- Inflammation
- Lack of movement—sitting, standing, lying down
- Low antioxidants
- Low arginine
- Low BH4
- Low estrogen
- Low glutathione
- Low oxygen
- Microbiome imbalance
- Mouth-breathing
- Overeating
- Oxidative stress (too many free radicals)
- Poor methylation
- Pollution
- Sinus congestion
- Sleep apnea
- Smoking
- Snoring
- Stress
- Tongue-tie

NOS3 and Your Other Dirty Genes

All your genes are continually affecting one another, as we've seen, but NOS3 is particularly affected by other dirty genes:

- A dirty MTHFR increases homocysteine, which in turn increases the biochemical ADMA, a component of blood plasma. In turn, ADMA uncouples NOS3, leading it to produce superoxide.

- A dirty GST or GPX decreases glutathione's ability to remove xeno-biotics and to eliminate hydrogen peroxide from your body. These harmful compounds reduce your levels of BH4, which then dirties your NOS3—and causes it to make superoxide.

- A dirty PEMT decreases your ability to maintain strong cell membranes, which leads to inflammation. Inflammation pulls your arginine away from NOS3, which leads to a dirty NOS3 and causes it to make superoxide.

- A dirty slow MAOA and/or a dirty COMT increases stress, which slows your methylation and raises homocysteine levels. As with a dirty MTHFR, this leads to increased ADMA, an uncoupled NOS3, and increased superoxide levels.

- A dirty fast MAOA may increase hydrogen peroxide levels and thus reduce your BH4. Low BH4 levels lead to NOS3 uncoupling and increased superoxide levels.

Your takeaway? If any one of those genes is dirty—or several are— you can be sure your NOS3 is dirty, too.

NOS3 and Dementia

If your Methylation Cycle isn't working well, you'll end up with elevated levels of homocysteine, which, as we've just seen, leads to increased levels of ADMA and then to a very dirty NOS3.

High ADMA levels are found in many conditions, including dementia. Interestingly, Alzheimer's disease is one of the main disorders associated with a dirty NOS3, while the second-leading cause of death for those with dementia is heart disease. It makes absolute sense. If the brain is inflamed and the Methylation Cycle is dysfunctional, NOS3 is going to be dirty. And a dirty NOS3 makes superoxide, which leads to cardiovascular issues.

This is yet another powerful reason to follow the Soak and Scrub to improve your Methylation Cycle, and then to Spot Clean your NOS3. Someone with mild dementia might even be able to reverse the condition, while someone with more severe dementia could slow its progression. Health professionals need to be much better versed in how significant methylation is. It seems to have its hand in everything.

Key Nutrients for a Healthy NOS3

You need both arginine and BH4 for your NOS3 to work properly, as we've seen. Arginine is the fuel in your tank, while BH4 is your starter. If you don't have both, your vehicle can't get going.

Making BH4 is a process that requires folate, magnesium, and zinc. You can't get BH4 directly from food: you have to support your MTHFR so that your body produces BH4. Please do *not* take BH4 supplements unless you were born with a rare form of BH4 deficiency. Research has shown that taking BH4 has no benefit if oxidative stress is present. It absolutely does not address the underlying problem. Remaining on the Clean Genes Protocol to protect your Methylation Cycle and maintain adequate glutathione levels is the best way to guarantee sufficient BH4.

Arginine, however, you can get from your diet:

Arginine: turkey breast, pork loin, chicken, pumpkin seeds, spirulina, dairy products (but stick to goat's or sheep's milk), chickpeas, lentils

Your NOS3 also needs the following nutrients to function:

Calcium: cheese, milk, and other dairy (but stick to goat's and sheep's milk products); dark leafy greens, bok choy, okra, broccoli, and green beans; almonds

Iron: seeds from squash and pumpkin; chicken liver; oysters, mussels, and clams; cashews, pine nuts, hazelnuts, and almonds; beef and lamb; white beans and lentils; dark leafy greens

Riboflavin/vitamin B$_2$: liver, lamb, mushrooms, spinach, almonds, wild salmon, eggs

Finally, your NOS3 needs plenty of oxygen—which you get from *breathing.* Now this seems obvious. What is *not* obvious is that many people struggle with sleep apnea, are mouth-breathers, have chronic sinus congestion, snore, unconsciously hold their breath, or breathe in a shallow way. Breathing is an absolutely essential, unconscious task that we do on average twenty thousand times a day. Do it wrong and it begins to create a significant problem. If you could implement only one change in your life to support your NOS3, I would—without hesitation—say to improve your breathing.

Exercise is healthy. We all know that. But did you know that moderate exercise actually helps support your NOS3 gene by making it work more efficiently?

Making the Most of NOS3

Rudy was committed to cleaning up his dirty NOS3. His high blood pressure was an early sign that he needed to make some changes. To create further incentive, I let him know that the erectile dysfunction he mentioned to me was another sign that his NOS3 was dirty.

The first recommendations I had for him were to reduce how much he was eating, get off his couch, move around at least a few times a day for twenty minutes each time, and switch his diet to a more gene-friendly regime instead of the standard American inflammatory diet. I knew that these three changes alone could make significant gains in restoring his dirty NOS3.

I also worked with Rudy on some deep-breathing exercises. I had him place his palm on his abdomen and then breathe all the way down to his belly, until he could feel his stomach pushing out when he inhaled. I asked him to breathe slowly and evenly, in and out through his nose, ten times, so he could feel what a huge difference it made to be fully oxygenated. I told him that throughout the day, I wanted him to check in and remind himself to breathe in this way—which would make a huge difference in cleaning up his NOS3 as well as in relieving stress.

Rudy was silent for a moment after I gave him my suggestions. "You know," he said thoughtfully, "my other doctors just told me to lose weight, exercise more, and take a blood pressure medication. I seriously thought about taking the medication, but I had no interest in changing my diet or in exercising."

Rudy paused for another moment. "The way you so clearly explained what's going on with my dirty NOS3—how it's linked to blood pressure, how it can lead to erectile dysfunction—makes me actually look forward to making some changes," he said. "It's not just 'Go lose weight' or 'Go exercise.' Now I know *why* I need to do these things. I guess it gives me a sense of clarity and purpose. Now that I know how these changes will clean up my NOS3, they don't feel like a hassle. It feels like I have the opportunity to make myself better. And *nobody* ever told me about the breathing!"

I was excited for Rudy. He was on his way, and he wouldn't need me harping on him about what he needed to do or stop doing; he'd gotten the message. A prescription lasts until your refills are expired. Education lasts an eternity.

Here are some other key things I let Rudy know about—steps that you can take, too, even before you begin your Soak and Scrub:

- Consume foods rich in natural arginine.

- Consume some foods that contain natural *nitrates*, which also support nitric oxide production: for example, arugula, bacon, beets, celery, and spinach.

- Keep building awareness of your breathing. You should be breathing at a nice, even pace—neither too fast nor too slow, and not erratically. You should also be breathing from your abdomen in deep, full breaths, rather than taking shallow breaths from your chest. Are you holding your breath at times? Do you snore? Consider breathing classes—for example, in either the Pranayama or the Buteyko breathing technique—or try some yoga or tai chi. If heart disease runs in your family or if your doctor has said it's a concern for you, improving your breathing is probably the best thing you could do to turn things around.

How the Clean Genes Protocol Supports Your NOS3

Diet. We'll make sure you get the right balance of nitrates and arginine, as well as all the other nutrients you need to support your NOS3. Eating healthy foods rather than inflammatory ones is a must; otherwise, you run the risk of running through all your available BH4. Avoid all foods and drinks that contain folic acid.

Chemicals. By reducing your chemical exposure, you'll keep your BH4 levels high. You also ensure healthy glutathione levels so that you can keep your Methylation Cycle happy. And remember, a happy Methylation Cycle is a happy NOS3.

Stress. Stress neurotransmitters increase the demands on your glutathione and methylation, as we've seen repeatedly. Both glutathione and methylation are needed to support your NOS3. Increased stress also increases your susceptibility to infections, which use up a ton of arginine and glutathione. That competition for needed biochemicals further reduces the amounts available for your NOS3, making it dirty. Most important, a common consequence of stress is rapid and shallow breathing. Insufficient oxygen is the fastest way to dirty your NOS3, and proper breathing is the fastest way to clean it up. In the Clean Genes Protocol, I'll help you practice breathing.

11

PEMT: Cell Membrane and Liver Problems

Marisol was a tall, elegant woman in her late fifties. She had been in menopause for about three years, and she was frustrated by the symptoms she had begun to develop.

"My triglycerides are very high," she told me. "My muscles ache, and so do my joints. I feel weak—I can barely lift a heavy pot up from the bottom shelf in the kitchen. And I've started to feel confused and foggy, as though I can barely concentrate. I forget things all the time. It's very frustrating!"

I had an idea what Marisol's problem might be, but I wanted to know more. "What happens when you eat fatty foods?" I asked her.

She stared at me. "How did you know to ask that? They don't sit right with me at all. I can feel the heaviness—right here." She placed a hand below her right ribcage.

"And Marisol," I said, "tell me a little bit about your diet. How often do you eat meat, liver, eggs, or fish?"

Marisol shook her head. "Hardly ever," she told me. "I'm not a vegetarian, but I mainly eat rice and beans or lentils for my protein, with maybe a little yogurt or cheese. Not much meat."

"Marisol," I told her, "it sounds as though you might have a dirty PEMT. That's the gene that makes phosphatidylcholine, a key ingredient of our cell membranes. But in order to make phosphatidylcholine, your body needs lots of choline—which you get from meat, liver, and eggs. There are *some* vegetable sources, but you may not be getting enough choline from the diet you've described."

Marisol looked surprised. "I thought too much meat wasn't good for you."

"*Too much* isn't," I said. "*Some* meat is good, or some fish and eggs. At the very least, you have to make sure you get enough vegetable sources of choline."

"But I've always eaten this way," Marisol said. "Why would I have problems now?"

I explained that, for many women, estrogen stimulates PEMT to serve as a backup, synthesizing phosphatidylcholine even when they're not eating enough choline. Before menopause, estrogen levels are higher and can often help compensate for a dietary shortage. During menopause, though, estrogen levels drop. This means that the PEMT gene doesn't function as well as it used to.

"All right," Marisol said slowly. "But what does that have to do with fatty foods?"

I told Marisol that a dirty PEMT is related to a syndrome called *fatty liver*. If you have that condition, your liver doesn't function well, partly because your PEMT isn't moving triglycerides out of your liver. A dirty PEMT can also contribute to muscle weakness and aches and pains, as well as brain fog.

"All these problems seem so separate," Marisol said finally. "I don't understand how they're all caused by the same thing."

I sympathized with Marisol's confusion. PEMT is a complicated gene with a wide variety of functions. A dirty PEMT works in subtle ways, too. You have to look at a number of different processes in your body to get the full picture. So let's get started—because PEMT is one of your most important genes, and supporting it can make a huge difference to your health.

Your PEMT in Action

Your PEMT is responsible for a number of tasks. By far its most important job is producing phosphatidylcholine, so let's start there.

Your cell membranes depend upon phosphatidylcholine. And those membranes are everywhere—surrounding every one of the 37.2 trillion cells that make up your amazing body. Each day, in an adult, over 220 billion cells die and have to be replaced. Every second, over 2.5 million red blood cells die and need to be replaced. PEMT is constantly helping to repair and regenerate the vast compilation of cells that is *you*, quietly behind the scenes.

Phosphatidylcholine keeps your cell membranes fluid and healthy, allowing them to function at their optimal level. If they become stiff, unhealthy, and nonfunctional, they can't move nutrients into your cells or transport harmful compounds out.

Compare a cell membrane to the outer walls of your home. You have doors and windows that open and shut. They lock tight to protect your belongings and your family. But the windows also open to let in fresh air while keeping out birds, flies, and mosquitoes. And your doors slide or swing open, with maybe even a specialized cat door so that the family pet can go in and out without bugging you. All of these doors and windows keep the warm air inside, conserving energy and ensuring a comfortable temperature.

Now imagine if your home had all its doors and windows removed. Your pets could come and go as they pleased—and so could your kids. Strangers could enter your home and take your belongings. Mice could invade your cupboards, leaving behind little black, smelly gifts. Your furnace would be working harder than ever to heat your home, but with all the doors and windows open, it would use up way too much energy for way too little result—and a very expensive bill.

Now apply that image to your cell membranes, which you can think of as the walls of your house. Your external cell walls protect your nucleus, which contains your DNA. Your internal walls surround and protect your mitochondria, the powerhouses that produce your body's energy source.

How can a cracked or leaky cell membrane protect your DNA? It can't. Think of all the environmental chemicals and infectious agents that will inevitably get in. And what will happen to your mitochondria without a healthy cell membrane? How can your mitochondria efficiently produce the energy each cell needs if their cell membranes are unhealthy? They can't.

In fact, without a membrane, your cell is dead. You can remove the nucleus from a cell and it will continue to live for a while. Remove the membrane, and the cell dies quickly.

We are an amazing collection of trillions of cells working in harmony. If you don't support your cell membranes, they can't support you.

How do we keep those membranes healthy?

Obviously, you want to eat health-promoting food and perhaps take some supplements as needed. But that's only the first step. You also have to *digest* the food and *absorb* its nutrients into your blood. You might be having trouble with this process without knowing it, due to taking antacids, eating processed foods, being stressed out, overeating, or drinking too much liquid during meals.

Ideally, though, your digestion is good, and the nutrients from your food and supplements are transported in your blood until they bind to a receptor or move inside your cells using protein channels. These receptors and protein channels are the "doors and windows" embedded inside each cell membrane.

Now you see why we want those membranes to be as healthy as they can possibly be, just as you want the walls and doors and windows of your house to function well. Without healthy cell membranes, some nutrients won't make it into the cells that need them, and you'll become functionally deficient in those nutrients—even if there are high levels of them in your blood. And without a high-functioning PEMT, you won't have healthy membranes.

PEMT: The Basics

Primary function of the PEMT gene

PEMT, with the help of your Methylation Cycle, helps your body produce *phosphatidylcholine,* a key biochemical that you need for several important roles:

- Phosphatidylcholine is the primary component of your cell membranes. Without enough of it, your cells are unable to properly absorb nutrients. You can develop malnutrition even if you're eating a healthy diet—in fact, even if you're overweight!

- You need extra phosphatidylcholine during pregnancy and breastfeeding. Children who are still growing also need extra phosphatidylcholine. Basically, whenever your body is making a lot of new cells, you need a lot of this vital substance.
- Phosphatidylcholine helps your bile flow smoothly out of your gallbladder to aid in digestion, thus keeping bacteria out of your small intestine.
- Phosphatidylcholine also helps package and move triglycerides, a type of fat, out of your liver. Without enough phosphatidylcholine, you can develop a condition known as *fatty liver.*
- In addition, phosphatidylcholine is essential for nerve function, muscle movement, and brain development.

PEMT also helps produce *choline* when you don't get enough of it from your diet. You need choline for a number of tasks:

- To support liver function, nerve function, muscle movement, energy levels, and metabolism.
- To make acetylcholine, a brain neurotransmitter important for learning and concentration.
- As a backup pathway for the Methylation Cycle when you don't have enough methylfolate (methylated vitamin B_9) or methylcobalamin (methylated B_{12}).

Effects of a dirty PEMT

With a dirty PEMT, you're unable to produce sufficient phosphatidylcholine. As a result, your cell membranes lose their integrity and the numerous bodily functions that depend upon phosphatidylcholine can't proceed smoothly.

Signs of a dirty PEMT

Common signs include fatigue, fatty liver, gallbladder disorders, inflammation, muscle pain, malnourishment (due to nutrients not being fully absorbed by damaged cell membranes), pregnancy complications, SIBO, elevated triglycerides, and muscle weakness.

Potential strengths of a dirty PEMT

With a dirty PEMT, you're better able to conserve choline to help with attention and focus. You also tend to have a better response to chemotherapy.

Meet Your Dirty PEMT

Your PEMT is the unsung hero of your genetic profile. Its work is vital to your well-being and health. But describing what your PEMT does is complicated, and it's even harder to explain how a dirty PEMT is connected to all the ways your body can go wrong.

So let's start with a few questions, over and above the ones you answered in Laundry List 1. The following factors can help you determine whether your PEMT is dirty and can give you a sense of all the different areas a dirty PEMT can affect:

☐ I have general pain everywhere—muscles, joints, all over.

☐ I'm a vegetarian or vegan.

☐ I had to have my gallbladder taken out.

☐ I've been told I have fatty liver, and/or someone in my family has fatty liver.

☐ I rarely eat leafy green vegetables.

☐ During pregnancy, my gallbladder acted up.

☐ I have SIBO.

☐ I've had genetic testing and know that I have the MTHFR C677T genetic polymorphism.

☐ I'm vitamin B_{12}–deficient.

☐ I'm intolerant of fatty foods.

☐ My estrogen levels are low.

☐ I take antacids.

☐ I have pain or discomfort in the upper right quadrant of my abdomen.

☐ My right shoulder is tight, by my scapula.

☐ I tend to have constipation.

☐ I tend to be itchy.

☐ I'm a postmenopausal woman.

The Life and Death of a Cell

The whole time you're supporting your cell membranes, millions of your cells are dying. That's natural and healthy: you want a certain number of cells to die every day—every minute, in fact—so that they can be replaced with brand-new cells. The cells of your gut wall, for example, are all entirely replaced within less than a week. Every week of your life, the old ones die while new ones replace them. Your red blood cells live for about four months; your white blood cells, for twenty days. Within about two or three weeks, all of your skin cells have died off and been replaced by new ones. It's the circle of life at its most basic level.

The process of a cell dying and being sloughed out of your body is called *apoptosis*, and as you can see, it's a *good* thing. The problem comes when too many cells are dying in relation to the new ones being born. Infection, inflammation, junk food, overexercise, and lack of nutrients all lead to apoptosis above and beyond the healthy balance that we're looking for. When our body takes a hit—an infection, a stressful week, a few nights without enough sleep—our cells are damaged. We need to be able to repair them very quickly or we'll start to have symptoms.

Suppose you keep eating a food to which you have a sensitivity. Or maybe you've developed some yeast overgrowth in your small intestine. As a result, your gut wall suffers some significant cell membrane damage, and apoptosis sets in. You need your small intestine cells to grow back healthy—but for that, they need phosphatidylcholine. Otherwise, your small intestine won't function well, you won't absorb nutrients, and you're likely to develop all the food sensitivities and related symptoms that go with leaky gut.

By the way, you don't want apoptosis happening too *slowly*, either. That could be a cause of cancer or allow cancerous cells to flourish. As with most things in the body, you want apoptosis to go neither too fast nor too slow, but just right.

PEMT and Muscle Pain

What if the cell membranes that are functioning poorly are those of your muscle cells? In fact, muscle cell membranes are quite fragile, and when they fail, that decline triggers inflammation. You start to get sore muscles, for no apparent reason. Prolonged phosphatidylcholine deficiency increases the severity of muscle cell membrane failure. In time, your muscles become not only painful, but weaker.

As you saw, that was one of Marisol's symptoms. Her dirty PEMT was failing to produce all the phosphatidylcholine she needed. As a result, she had aches, pains, and muscle weakness.

Health Conditions Related to a Dirty PEMT

As we've seen in this chapter, a wide range of conditions are impacted by a dirty PEMT. They include the following:

- Birth defects
- Breast cancer
- Depression
- Fatigue
- Fatty liver
- Gallstones
- Liver damage
- Muscle damage
- Nutrient deficiency inside your cells
- SIBO

Key Nutrients for a Healthy PEMT

Your PEMT makes from 15 to 30 percent of your phosphatidylcholine, and in a pinch, it can supply more—but that's when it starts to stagger under the burden of its job. To avoid that dirty situation, you need a good supply of dietary choline so that other genes can use it to make phosphatidylcholine:

Choline: liver, eggs, fish, chicken, red meat

As you can see, getting dietary choline is tough for vegetarians and vegans. It's also tough for "carbotarians," people who feast solely or primarily on carbohydrates. People who don't eat meat or eggs are at high risk for choline deficiency, which means they'll run short on important compounds made from choline, including phosphatidylcholine. However it comes about, a low-choline diet puts people at risk for fatty liver, liver cell death, and muscle damage. I was reasonably sure that such a diet was largely responsible for Marisol's symptoms.

Young women have an extra capacity to make choline, because, as we saw earlier, estrogen stimulates their PEMT. That makes them less dependent on dietary choline: their PEMT can fill in the gap. This makes sense when you think that young women are the ones who get pregnant and breastfeed. They need a whole lot of choline to bear and nurture children, so nature has arranged backup.

But even young women suffer from a low-choline diet if they were born with a certain type of dirty PEMT—the type that doesn't respond to estrogen. And that particular SNP, it turns out, is quite common—and a significant health hazard. Some studies have shown that the less choline a woman consumes, the higher her risk of breast cancer.

If you're not a vegetarian or vegan, make sure to get enough animal protein in your diet—not too much (we don't want too much histidine, as we saw in chapter 7, or too much tyrosine, as we saw in chapter 6), but not too little. And if you're a committed vegetarian or vegan, try some alternative sources of choline:

- Asparagus
- Beets
- Broccoli
- Brussels sprouts
- Cauliflower
- Flaxseed
- Green peas
- Lentils
- Mung beans
- Pinto beans
- Quinoa
- Shitake mushrooms
- Spinach

In the Clean Genes Protocol, I'll also suggest ways to supplement with choline and/or phosphatidylcholine, especially if you're a vegan or vegetarian. But as always, I want you to start with diet and lifestyle solutions first.

Who Is Most at Risk for Choline Deficiency?

Pregnant and breastfeeding women. As this chapter stresses, choline is needed to make phosphatidylcholine, which is used to make cell membranes. Since women who are pregnant or breastfeeding are making a *lot* of new cells, they need a *lot* of choline.

Children. As children grow, they create lots of new cells every day, so they need lots of phosphatidylcholine, too. Without enough choline in their diet, they could be at risk of a phosphatidylcholine deficiency.

Vegans and vegetarians. It's hard to get enough choline from vegetable sources, although vegetarians who eat eggs have some advantage. (See above for a list of vegetable sources of choline.)

People fasting inappropriately. If you choose to go for more than forty-eight hours without food, consider supplementing with choline and/or phosphatidylcholine. It isn't a healthy choice, however. I'd rather you gave your body the nourishment it needs.

People on a low-protein diet. Without consuming substantial protein, it's hard to get enough choline. A *high*-protein diet isn't the solution, though. We're always looking for the right balance.

Postmenopausal women. High levels of estrogen trigger a clean PEMT to make phosphatidylcholine—but after menopause, estrogen levels drop. Even if you were born with a clean PEMT, menopause could make it dirty—unless you get enough choline in your diet. Of course, if you were born with a certain type of dirty PEMT, you won't have this "estrogen advantage" even before menopause.

Men. Without high levels of estrogen to trigger the backup action of their PEMT, men must be sure to eat a high-choline diet.

People who are low in folate, or with dirty genes in the folate pathway (MTHFR or MTHFD1). The amounts of folate and choline in your body are related. Too little folate means that your body uses way more choline—and you could end up deficient.

People with a dirty PEMT. If your PEMT is dirty, it doesn't respond to estrogen—so you need more choline in your diet.

PEMT, Methylfolate, and the Methylation Cycle

You need a lot of choline to make phosphatidylcholine and keep your cell membranes in good shape. But here's another factor: the more methylfolate you have available, the less choline you need, and vice versa.

Why? Because both methylfolate and choline support the Methylation Cycle. So if you have a lot of methylfolate available, the Methylation Cycle doesn't need to draw on your choline supply. But if you're short on methylfolate—as you might be if you have a dirty MTHFR or MTHFD1 (another gene in the folate pathway)—your Methylation Cycle will use the choline pathway. And now you're potentially short on choline, risking a dirty PEMT as well as several other problems.

Your best protection? Make sure you have all the methylfolate you need *and* all the choline you need.

But monitoring those biochemicals is no guarantee of gene health. Even if you have enough methylfolate, your Methylation Cycle might be disrupted at another point—and then your PEMT will get dirty.

Why? Because your PEMT needs SAMe, which we met in chapter 5. And in order to have enough SAMe, you need a highly functional Methylation Cycle.

About 70 percent of your SAMe is used by your PEMT enzyme to support the production of cell membranes, leaving only 30 percent for the other two hundred processes that depend on methylation. That's why extra demands on the Methylation Cycle—stress, excess histamine, insufficient B$_{12}$ (from being a vegan/vegetarian), chronic illness—are so hard for your body to handle. The amount of SAMe needed is already great—and now you're using up even more.

Likewise, if your Methylation Cycle isn't working properly in the first place, your supply of SAMe is low. Then your PEMT gets dirty, your cell membranes suffer, and your entire body feels the impact. Once again, the Methylation Cycle turns out to be of *prime* importance.

What Makes PEMT Dirty?

- Not enough choline in your diet
- Not enough methylfolate in your diet
- Not enough SAMe
- A disrupted Methylation Cycle
- A dirty MTHFR
- Insufficient estrogen—if you're a postmenopausal woman or a man
- A born-dirty PEMT SNP that doesn't respond to estrogen

PEMT and the Digestive Role

The gallbladder, bile flow, and liver are all impacted by PEMT. Let's take a look.

PEMT, Gallstones, and SIBO

Besides cell membrane health, PEMT-triggered phosphatidylcholine is also vital for bile flow. Your gallbladder makes bile to aid in digestion and, with its antimicrobial properties, to protect you from small intestine bacterial overgrowth, or SIBO. If your phosphatidylcholine levels are too low, your bile flows sluggishly. Your gallbladder then begins to malfunction, potentially leading to gallstones, fat malabsorption, nutrient deficiency, SIBO, and chemical sensitivity. Gallstones are especially prevalent in pregnant women because of the high demands for phosphatidylcholine.

PEMT and Fatty Liver

There are many different causes for fatty liver, which happens to be the fastest-growing condition in the world. Researchers are finding that fatty liver is on the rise for a number of reasons such as high fructose corn syrup, metabolic syndrome, obesity, and medications. Scientists have recently discovered that a dirty PEMT is one such cause: it promotes fatty liver. Marisol, whom we met at the beginning of this chapter, didn't have fatty liver yet, but symptoms showed that she was well on her way. So are many young women born with a dirty PEMT and not getting enough choline.

How does a dirty PEMT contribute to fatty liver? There are two ways, both tied to PEMT's role in producing phosphatidylcholine.

First, if your PEMT is dirty and doesn't trigger enough phosphatidylcholine production, you'll have triglyceride trouble. You need phosphatidylcholine to move your triglycerides out of your liver, which is done via "very low-density lipids" (VLDLs) secreted by the liver. If you're deficient in choline (and thus in phosphatidylcholine), your liver doesn't make enough VLDLs, and so your triglycerides build up. Pretty soon, you've got way too much fat remaining in your liver instead of moving out into your bloodstream where it will be eventually transported and used as fuel by your mitochondria.

Second, phosphatidylcholine is needed to make cell membranes. As phosphatidylcholine levels drop, your mitochondria become less able to burn fuel. When you can't burn fat as fuel, you store it in your cells, where it causes oxidative stress. This stress further damages your mitochondria, which then burn even less fuel. Now you've got a vicious cycle that can't repair itself until your cell membranes are healthy—and meanwhile, your body is storing fuel as fat.

PEMT, Pregnancy, and Breastfeeding

Sadly, most pregnant and breastfeeding women are deficient in choline. And shockingly, most infant formulas provide little to no choline. Please, if you're planning to get pregnant, see a reliable physician or nutritionist who can make sure you're getting enough choline in your diet. Research shows that women with a diet low in choline have 2.4 times greater risk of having babies with neural tube defects, such as spina bifida. The highest blood choline levels were associated with the lowest risk. I recommend that women get 900 milligrams of choline a day while pregnant and breastfeeding.

After the baby is born, breastfeeding demands even more of your choline levels, because this nutrient is secreted into breast milk at high concentrations to support the development of your infant's brain, liver, and cell membranes. Put another way, the developing baby needs that extra choline for three primary reasons:

- To promote cognition
- To support methylation
- To make cell membranes

Researchers have found that mothers who consumed more choline during pregnancy have babies with improved memory and learning abilities, while mothers with a lower-choline diet during pregnancy tend to have children with decreased memory and more learning disabilities. A number of studies have found that *most* pregnant women in the United States are choline-deficient, so please work with a naturopathic physician or integrative/functional medicine doctor to make sure that you and your baby are getting all the choline and methylfolate you need.

Making the Most of PEMT

Marisol was concerned to realize the full extent of her health problems, but relieved to learn that she could turn her situation around. I urged her to start with diet. Since she was still reluctant to rely too much on meat, I suggested that she start making egg salads, which are a mainstay of my own diet. (I like egg salads made with soy-free mayonnaise, pickle relish, salt, pepper, and chopped romaine, perhaps served on gluten-free toast.) I also recommended deviled eggs, which I personally prefer to plain old scrambled or fried eggs.

Here are some more suggestions for supporting a dirty PEMT that you can begin immediately, even before you start the Clean Genes Protocol:

- Make sure to eat some high-choline foods every day. It doesn't matter whether you choose meat or vegetable sources—just make sure you're getting enough choline. I was surprised when I evaluated my own diet and found out that I wasn't getting enough choline. My liver reminded me when I began feeling queasy. If you get enough choline in your diet, then having a born-dirty PEMT gene is much less of an issue.

- Eat in moderation. This is good for everyone, but it's particularly important for you, because those extra proteins, carbs, fats, and sugars burden an already stressed liver. The solution is simple. Stop eating when you feel about 80 percent full. Give it fifteen minutes after you've cleared your plate, and the sense of satiety will come. Your liver will thank you.

- Control your stress. We all need stress relief, but it's especially important for you, because stress burns through choline like nobody's business. Keeping stress in bounds gives your dirty PEMT a chance to catch up.

- Eat leafy green vegetables. The lower you are in methylfolate, the more choline you have to use to support your Methylation Cycle.

- Make sure you *absorb* and *digest* your protein. This requires chewing thoroughly, eating while calm, not drinking more than eight ounces of liquid during meals, not taking antacids, and not driving while eating.

- Reduce your intake of refined carbs. That means cutting out the chips and crackers! Go for proteins, healthy fats, and complex carbs, such as snap peas, baby carrots, and hummus.

- Wash your hands—with natural soaps, *not* antibacterials—before eating or after spending time in highly public places such as airplanes, hospitals, schools, offices, and sporting facilities. This will help you reduce infections from bacteria and viruses, which will ease the burden on your system and reduce your need for extra choline.

- Cook with avocado oil, sunflower oil, and/or ghee to reduce fatty acid oxidation. Don't cook with coconut oil or olive oil, as they have low smoke points. Also, be sure to turn on your stove fan while you cook.

- Love your liver. Because 85 percent of all methylation reactions occur in the liver, easing the strain on that organ will protect your phosphatidylcholine levels, and your liver as well. Limit your alcohol intake and cut out preservatives and all unnecessary medications (with your doctor's permission).

- If your gallbladder is sluggish, consider working with a professional who specializes in visceral manipulation. These practitioners can gently work to manually drain your gallbladder. I personally had this done, and it worked beautifully. The manipulation is fast and effective and allows immediate relief while you change your diet and lifestyle.

How the Clean Genes Protocol Supports Your PEMT

Diet. We'll make sure you get enough choline and methylfolate in your diet, while removing the foods and beverages most likely to stress your liver: high-fat foods, foods with preservatives, and alcohol.

Chemicals. By helping you to heal your leaky gut and strengthen your digestion, we'll protect you from pathogenic bacteria and infections that stress your system, eat up choline, and burden your Methylation Cycle.

Stress. Both physical and emotional stress burn through choline. By focusing on stress reduction and stress relief, we'll help you conserve your choline and support your PEMT.

PART III

YOUR CLEAN GENES PROTOCOL

12

Soak and Scrub:
Your First Two Weeks

When you get home from the gym or you're just finished with a tough game of soccer, your clothes are sweaty and stinky, and so are you. What's the first thing you do? You take off all your clothes and jump in the shower, soap up to wash off the sweat, and then dry yourself off. Only then do you put on a fresh set of clothes.

You don't see it, but your genes are dirty. We humans tend to fix what we see and ignore what we don't. But your headaches, rashes, weight gain, and insomnia are all results of one or more dirty genes. Popping a pill or a supplement doesn't fix anything. It just tells your genes to shut up.

The problem is, they don't listen. They have set instructions, and that's all they know. They depend on you to understand how they work so that together you can achieve optimal genetic efficiency.

Now, you and I have journeyed this far together. You've heard me say over and over again that we have to address the foundation of the dirty genes problem. We've arrived at that point now. Here, in part III of this book, you'll discover what your genes need to function at their best. And for the next two weeks, that's exactly what you're going to give them.

This program is designed to give your genes all the support they require.

So before you do *anything*, I want you to stop and ask yourself:

Is this food, supplement, or activity going to support my genes or make them work harder?

Remember, if your genes have to work harder, then you'll likely get symptoms you don't want.

Think of it this way. When you're working, you get tired. When you go on vacation, you get recharged and are ready to tackle any problem you face when you return to work. For the next two weeks, you're going to give your genes a vacation. No one can work twenty-four hours a day every single week of the year without consequences. Not even your genes. Give them a break.

Once you understand the consequences of your actions, you'll make better decisions. Your genes will thank you for it—and you'll start to feel stronger, leaner, and more energized.

Most important, you'll feel empowered knowing this amazing information. You'll finally have many answers to nagging issues you've wrestled with your entire life.

Soak and Scrub: Week 1

You might already be following some of the recommendations I provided earlier to clean up your genes. If so, and you're having success, fantastic! Keep on implementing those changes. Simply add the new recommendations below to further enhance cleaning up your genes. If you're trying some of those earlier recommendations and not having success, stop them and follow what you find outlined below for the Soak and Scrub. If you haven't tried them at all, no worries. Now is the time to do so.

In the pages that follow, the recommendations are grouped according to the diet and lifestyle elements covered by Soak and Scrub:

- Food
- Supplements
- Detoxification
- Sleep
- Stress relief

Let's turn to the biggie, food, and get you started!

Food

As Hippocrates wisely said, "Let food be thy medicine."

But here's the thing. Just as each of us might need different types of medicine, we each need different types of food.

We all know that while a certain medicine might help one person, it hinders another by generating significant side effects. The exact same thing is true of food. While fermented foods might be great for me, replenishing my microbiome and healing my leaky gut, you might have a dirty DAO that can't handle the extra bacteria. Maybe you can tolerate small amounts of gluten, while I can't. My son Mathew gets a runny nose, irritability, and earaches from cow's milk dairy, while Theo responds to the same thing with frequent eye blinking and constant clearing of his throat. Tasman, meanwhile, can eat cow's milk dairy with no symptoms. We are all different, and our reactions to food reflect that.

We are also each changing, all the time. Maybe a food that worked for you last year gives you massive symptoms this year—or perhaps it's the other way around. Different genes become dirty or get clean, and since all your genes are always talking to one another, your whole biochemistry is constantly changing in response.

That's why *tuning in to how you feel* is the cornerstone of your life on the Clean Genes Protocol. The goal of this book is to teach you how to live as optimally as possible so that you can reach—and maintain—your genetic potential. You can do that best by learning how to tune in to yourself, so that you always know what to eat, and when.

Tune In to Your Body and Emotions

Let's face it. Some days you feel like eating out and other days you feel like just a salad is enough. That's life.

Over the years, I've learned to tune in to how I'm feeling mentally, physically, and emotionally. That's how I figure out what I'm going to eat. If you're like most people, when you're stressed out, you go for carbohydrates and suck down calorie-dense foods to give yourself a feel-good dopamine rush. The issue? It doesn't last. So you do it again and again, only to gain weight and feel irritated with yourself about the choices you made—but you can't help it. It gives you a lift.

I get it.

That's why the first step to tuning in is recognizing the difference between *cravings* and *hunger*. Easier said than done, of course.

It's especially hard to understand the difference between cravings and hunger when you're in the craving state. A craving is a feeling that you *want* to eat something specific, while hunger is that gurgling, empty feeling in your gut that you *need* to eat.

We all have dirty genes that give us cravings. The dirtier our genes, the more cravings we have. That's one piece of terrific news: remaining on the Clean Genes Protocol will greatly reduce your cravings.

At the beginning, though, your cravings will be screaming at you, "Give in! Fail! This is too hard!" Don't listen to them. I've learned ways to shut them up, and I'll share those with you.

For now, though, simply try to switch your mindset from *craving* food to actually *being hungry*. Simply ask yourself, "Do I need to eat or do I want to eat?" If you succeed in this, your genes will thank you for the rest of your life. But meanwhile ...

Banish Regrets

We're all human. We all have cravings. We all enjoy food that tastes exceptional and tickles our taste buds. We all deserve to experience amazing food and that sense of warmth that comes over us after a great meal.

And so, yes, there will be days and nights when you go to town and overeat and likely consume foods you know aren't ideal for you. That's fine! I do it, too. The next time you do this, enjoy it. Go for it. Don't beat yourself up over it the next day. No guilt trips. You made a decision, you enjoyed the meal—now the next day is a new day. To paraphrase my colleague Dr. Sachin Patel, "Your decision to be healthy starts with your next bite." How beautiful is that? Just go back to cleaning up your genes and supporting the ones that need some extra help. The cool thing is that now you know how to do that.

Just promise me that the next time you overindulge or eat a food you know you shouldn't, you'll *enjoy* it. Feeling guilty or regretful will only make your genes more dirty.

Think of it like this. You invited a load of great friends over, hired a band, had a bouncy house for the kids, and cooked for twenty people. People were laughing, dancing, and having an exceptional time. On their way out, each of your friends gave you a huge hug and said, "Wow—awesome party! Thank you!" and you felt great, too.

The next morning you wake up, tired and heavy, and with bleary eyes, and you face a huge mess. Stains on the carpet, dishes everywhere, dogs on the table eating scraps, the yard trashed from the kids. You smile, remembering the great evening. You turn on your favorite CD and start cleaning up the mess, one room at a time. No regrets!

Plan Your Meals

Most of us just follow our cravings. Instead, I want you to tune in, listen to your body—and *plan*.

Why?

Food should be restorative and nurturing. Many of us don't think of food this way. We think instead, "Oh, what do I feel like? Something salty, sweet, fatty, chewy, crunchy?" or "Dang it. I've so much work to do, but I have to eat. I'm gonna run out and grab something and eat it really quick."

I want you to switch to a different mindset. Food is beautiful. Food is your fuel. It's not a nuisance to get out of the way. Think of it this way: "Hmm ... I have a big day ahead. I'm doing a presentation at work, taking the kids out to play soccer, and then we're off to watch a movie. What do I need to eat in order to succeed today? Protein to help me think. Some complex carbs to keep me going while I run around on the soccer field, and a light salad before I sit on my butt watching a movie."

Making that kind of plan takes awareness of what you'll be doing and what your body will need to succeed. Tune in to your body and make your food plans based on the following factors:

- **Your activity level, including both mental and physical activity**. More mental activity requires more protein for sustained acuity, while more physical activity requires protein, healthy fats, and carbs for sustained energy.

- **Your emotions—happy, sad, mad, enthusiastic, bored.** Happy and enthusiastic moods require less food, as does being bored. Extreme emotions—such as sadness and anger—might require more or less food, depending on the situation. However, being bored, sad, or mad typically drives *cravings* for food—not actual *hunger.*

- **Your symptoms (or lack of symptoms).** Do you have a headache? Feel heavy? Have brain fog? Can't sleep? Have no energy? Feel stressed? Or do you feel great—clear-headed, energetic, sharp? Feeling great, clear, and sharp requires less food. You've got something good going, so don't mess it up with extra food. Headache, lack of energy, trouble sleeping, stress, and brain fog may be due to inappropriate food choices, and you're experiencing the consequences. On the other hand, if you haven't eaten for a long time, those symptoms may be signs that you need to eat or hydrate. Tune in so that you can make the right judgment call. *Next time,* you'll make the right decision.

- **Your genes.** Which ones need cleaning? Which need additional support? Food fuels your genes. It's your job to deliver what they need so that they can perform at their best—and you can too. Give them garbage, and that's what they'll deliver in return. Give them quality nutrition and time to utilize it, and they'll do their best for you so you can succeed.

Track Your Meals

Knowing how a food is helping or hindering you is important. Tuning in definitely helps. Taking it a step further and tracking what you eat in a food journal is even better. This way, when symptoms occur, you can look back and see what you ate and deduce what might be contributing to those symptoms. Tracking also helps you see the big picture of your diet—the amount of protein, carbs, and fat you're eating, for example, and what times of day you typically eat more and less.

When I started tracking what I ate, I discovered that I was eating way more carbohydrates than I'd thought, which was contributing to my sleepiness, brain fog, and weight gain. I also discovered reasons why I wasn't sleeping at night—eating too much, too late, and too much protein.

I prefer to use an app like CRON-O-Meter, which is easy to use. There are also other programs (see the Resources section) that can help you.

Eating to Clean Your Genes: The Basics

- The easiest way to keep your genes clean is not to make them dirty. By eating organically, you reduce how much work your genes have to do. Organically grown foods also have more nutritional content compared to nonorganically grown.

- Cost can be a limiting factor with organically grown foods. If this is the case in your household, buy organic for only the worst offenders—those fruits and vegetables that, when conventionally farmed, are highest in toxins. Check out the lists of "dirty" and "clean" foods—conventionally farmed foods most and least contaminated by industrial chemicals—maintained by the Environmental Working Group (www.ewg.org). Every year, the EWG determines the "Dirty Dozen" and the "Clean Fifteen." Typically, these lists don't change much. Commonly, the following fruits and vegetables are those to avoid if not buying organic: strawberries, apples, nectarines, peaches, pears, celery, grapes, cherries, spinach, bell peppers, tomatoes, cherry tomatoes, cucumbers, hot peppers, and kale.

- If you aren't hungry, don't eat. Now, of course there are some exceptions—such as if you know you're facing a long stretch of time when you won't be able to eat—but for the most part, this injunction generally applies.

- Eat until you're 80 percent full—then *stop*.

- Eat a maximum of three meals a day. Ideally eliminate snacking if you can. If not quite yet, at least limit it.

- If you find yourself snacking, consider the following common reasons:
  - You're experiencing a craving rather than true hunger. Don't give in to the craving. Be strong. Ask yourself, "Do I want to eat or do I need to eat?"
  - You've got a bad habit—reaching for food between meals—that you need to break. Break it!

—Your fuel-burning functions aren't working properly.

—You're eating, but you're not absorbing your nutrients.

—You're not eating health-promoting foods, which means your body never feels satisfied. On top of that, eating poorly creates inflammation and even malnutrition.

- Fast for twelve to sixteen hours daily. This is easily accomplished if you stop eating at 7 P.M. and then wake up to breakfast at 7 a.m. If you finish dinner at 7 P.M. and don't eat again until 11 A.M., that's sixteen hours. Personally, I feel the best most days when I stop eating at 7:30 P.M. and break my fast at 11 A.M. or noon. If I'm leaving out the door to a conference or have a presentation in the morning, I'll have breakfast earlier. At the first sign of a blood sugar drop (slower thinking), I make sure to eat.

- Chew, chew, chew. Take a bite of food. Put your utensil down. Chew completely. Enjoy the flavor. Savor it. Swallow. Repeat. I would say 99 percent of us don't chew thoroughly, allowing time to appreciate each bite—but this alone will greatly decrease how much food you eat and increase how good you feel.

- Limit drinking during meals. Have one glass of filtered water, goat's milk, almond milk, tea, or—every so often—wine, but never any more than *one* glass. Don't dilute your digestive enzymes. Doing so limits your ability to absorb what you're eating. Trust me on this.

- Don't drink cold beverages during meals. Room temperature or warmer is best. Cold temperatures require your body to warm it up thereby depleting you of energy. By drinking water that is cool but not cold, you are conserving energy. Sorry, no, you won't lose weight by drinking cold water. Cold water may also cause stomach cramping and colic, especially if exercising.

- There's no "mostly" gluten-free. If you have even a bite or two of gluten, that potentially sets off the same biochemical reaction as eating a loaf of bread. Why? Because your immune system responds via antibodies, and the antibodies are triggered by even tiny amounts of food. So 99 percent gluten-free is the same as 0 percent. Either you're 100 percent gluten-free or you aren't.

- If you have a fever, don't eat. Just hydrate with electrolytes. Of course, if you have a prolonged or high fever, you need a health professional to assist you.

- Drinking fruit juice is like drinking soda. It's pure sugar. Limit it. I avoid both juice and soda 100 percent. It took me years to adapt to this, but I accomplished it; and I feel so much better without either. Drinking fruit juice quenches a craving. It also dirties your genes right up.

- Juicing at home is great, but be sure to juice vegetables and not fruit. Ideally, use a Blendtec or Vitamix and blend whole vegetables and herbs to get all the nutrients and fiber. Organic produce is preferred, of course, so you don't load yourself up with herbicides and pesticides. That'd be one dirty smoothie.

Make Wise Food Choices

Getting your food choices dialed in is going to help you make great strides in cleaning up your dirty genes. The key is knowing what you can and cannot eat. I've made that much easier for you: in the next chapter, you'll find amazing recipes for breakfast, lunch, and dinner.

Before choosing your recipes, complete Laundry List 1 in chapter 4 (if you haven't already) to discover which of your genes are dirty. Armed with this insight, select the recipes that will help you clean those genes. Make yourself a shopping list and get started!

Here are some additional guidelines for eating to clean your genes:

- **Avoid foods that are stocked in the middle aisles of the grocery store, foods with ingredients you can't pronounce, and foods that are white:**
  —Soda, diet as well as regular
  —Fast food
  —Anything that contains folic acid (which is *everywhere* in processed foods)
  —Ready-to-eat frozen dinners or ready-to-eat packaged foods
  —Cold breakfast cereals (oatmeal and other hot, gluten-free cereals are okay)

—Granola

—Chips

—Snacks, including crackers, trail mix, granola bars, energy bars, and anything else that doesn't create a full meal

—Candy

—Ice cream

—Energy bars

—Juice

—Unfiltered water

—Gluten

—Soy

—Dairy

—Alcohol

- **Focus on foods stocked on the perimeter of the grocery store, foods without added ingredients, and foods that the planet provides you naturally:**

  —Filtered water

  —Loads of fresh vegetables

  —Some fresh fruits—no more than three daily servings; best eaten in the morning or afternoon and not at night

  —Eggs, organic or free-range

  —Free-range meat, ideally from the local rancher or local butcher—grass-fed beef, lamb, bison, venison

  —Fish and shellfish—wild, fresh-caught

  —Nuts and seeds

  —Sprouts of all types: beans, grains, seeds, and nuts

  —Wild rice

  —Quinoa

  —Fresh deli ready-made foods from a natural food co-op—chili, soups, salads, entrées—all of which are excellent when you're busy. Be sure to read the ingredients and avoid the foods that aren't health-promoting for you and your genes.

- **Individualize your meals.** Specifically tailor each meal to how you scored on Laundry List 1. A full guide of what and when you should eat is provided in the next chapter.

- **Cook or steam foods fresh.** Avoid frozen foods and leftovers. Leftovers are especially problematic if you're struggling with a dirty DAO.

- **Digest your food.** Thirty percent of your stomach acid is released in response to *preparing* to eat: observing your food, smelling it, and looking forward to eating it. Take the time to do all three of these things. Food should be nourishing *all* of you—mind, body, spirit—and genes. A meal shouldn't be something that you rush through and "get over with." To see what I mean, think *lemon* right now. Did you feel the flow of saliva in your mouth? That's a sign that your digestion is ready to rock. Anticipating and "presavoring" your food in this way will make it taste better—and you'll eat less, burn food way more efficiently, and support a healthy metabolism.

- **Use your stove exhaust fan when cooking.** I know it's noisy, but the oil smoke is toxic. The less you breathe in, the better.

- **Cook with high–smoke point oils only.** Ghee, avocado oil, sunflower oil, and safflower oil are best for cooking or baking. Olive oil, coconut oil, flaxseed oil, and walnut oil are great for salads.

Supplements

While it's our goal to get all our nutrients from food, that isn't always possible. Nutrient density in food is lost for various reasons. For example, soils are often depleted; in addition, transportation, extremes in temperature, the cooking process, and time on the shelf all degrade nutrients. We're also exposed to chemicals and stressors each day that burn through our much-needed nutrients. For all these reasons, sometimes we need to supplement.

Here are some basic principles regarding supplementation that most people don't follow—but that will make a huge difference in your health if you do!

Choose the Form *of Supplement That's Best for You*

When I talk about the *form* of a supplement, I'm referring not to the nutrient it contains, but the means by which that nutrient is delivered.

The easiest form of supplement to absorb is liposomal (delivered via microscopic fat balls in liquid). The hardest is tablet. Here's the order, from easiest to hardest to absorb:

Liposomal (liquid) > lozenge > powder > chewable > capsule > tablet

When you want to regulate how big your dose should be, the order is the same. Liquids are easiest when you want to fine-tune a dose; tablets are hardest. It's easy to take ¼ teaspoon of something, but it can be tough trying to hack a tablet into equal fourths!

Here are some other things for you to consider when deciding which form of supplement to take:

- If you're sensitive to supplementation, start a liposomal form with just a single drop. If you tolerate supplements well, start with a ¼ teaspoon. The nutrients, in this liquid form, are delivered right into your cells.

- Taking a lozenge is great as well, not only for absorption but also for regulating the amount you're getting. If you place the lozenge in your mouth and tune in, you can often experience the action of that supplement quite quickly—sometimes within minutes. If you're feeling better, or if you don't feel anything, great; let it continue dissolving. If you're feeling worse, though, take it out. You might not do well with this particular product, or you might want to cut the lozenge into quarters or halves to regulate your dose.

- Powders are also great, because you can easily adjust the dose. Taste is an issue here for some, but there are many great-tasting powders. The ones that don't taste so great can be mixed with an ounce of juice or little bit of applesauce.

- Chewables can often be cut into halves or quarters if you need to adjust the dose.

- Capsules are handy in that they mask the taste and protect the nutrients from air and water. Most high-quality capsules dissolve quite well in the stomach or small intestine. Oftentimes people want to open capsules and sprinkle the contents directly into their mouth or in food or water. This is fine as long as you check with the manufacturer or your practitioner. Some capsules should not be opened in this way—for example, betaine HCl, as it is highly acidic and can burn.

- Tablets aren't generally useful unless they were designed as a sustained-release form of a supplement such as niacin. If you have low stomach acid or are taking antacids, tablets may not dissolve well in your gut. Then you just end up with expensive stool. A park ranger once told a professor of mine about how many vitamins he sees in the outhouse—tablets that made it through whole! X-rays also often show undissolved tablets in people's digestive systems. Tablets may be generally cheaper to make (and buy), but in the end they're expensive, because you're wasting time and money.

Don't Feel Tied to "Suggested Use"

The "suggested use" directions you see on a supplement bottle are just that: suggested only. Follow your health professional's prescription—or your own sense of what works for you. I always believe in starting small, to see how a supplement affects you. If the "suggested use" is four capsules a day, start out with only one capsule a day. The amount of nutrients you're getting would then be one-fourth of what's shown on the supplement facts label.

Take One Supplement at a Time

I understand the impulse to load up on lots of new supplements. I get excited too! When I first began working with patients, I knew which supplements could help them, and I often recommended several at a time to get started. When it worked, it was awesome. When it didn't, it was a nightmare, because we had no idea which supplement had caused the problem. I've learned the hard way to try just one supplement at a time. Take it for a few days and see how it's working for you. Only then, once you see either a benefit or no change, should you add in another supplement. (I say "no change" because the supplement might not have had time to work yet—but at least it isn't causing you any harm!)

Know Your Body and Know Your Supplement

Before you swallow any supplement, you need to understand what that supplement is designed to do. What is its purpose? To raise serotonin and slow your fast MAOA? To clear out dopamine and support a slow COMT?

When you've purchased a particular supplement and are ready to try your first dose, take a moment to notice how you're feeling. Be aware. Tune in.

Then notice how you're feeling. Some supplements act within minutes, such as NADH. Others may take thirty minutes (such as acetyl-L-carnitine) or twenty-four hours (such as ashwagandha) to act. Again, be aware and tune in. Did the supplement do what you thought it was going to do? Is your slow COMT working faster now? Did you slow down your fast MAOA?

You're responsible for your own health—but more important, you know your body better than anyone else ever could. Only by listening to what your body tells you can you decide which supplements you need—and when you don't need them anymore.

Follow the Pulse Method

The Pulse Method is my approach to figuring out how much of a supplement you need—and at what point you should increase the dose, decrease it, or cut the supplement out altogether. It is *vital* for you to understand and use this method. Otherwise, you'll end up taking supplements long after you need them and maybe even after they've begun to do you harm. If your body is deficient in something and you fill up that deficiency with a supplement, chances are you won't need that supplement anymore. If you keep taking it, you might well end up with a surplus, pushing your system into a new extreme with new and unpleasant symptoms.

So here is the Pulse Method, your guide to taking supplements:

As the diagram shows, the moment you feel great is the moment you should *stop or reduce* a supplemental nutrient. First lower your dose—then keep lowering it, until you're down to little or no dose at all. If you feel bad again over time, you can increase the dose gradually. But if you start getting *different* symptoms, that might mean you're taking too much.

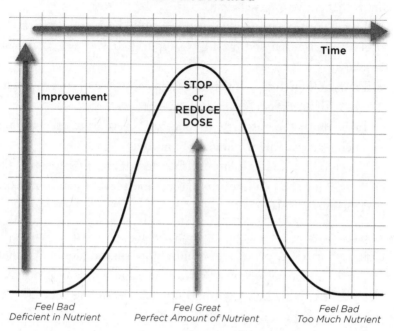

The Pulse Method

Time

Improvement

STOP or REDUCE DOSE

Feel Bad
Deficient in Nutrient

Feel Great
Perfect Amount of Nutrient

Feel Bad
Too Much Nutrient

For all the supplement recommendations below, you need to be using the Pulse Method. These supplements are powerful and highly effective. Use them *only* when you need them. Stop or lower the amount when you feel great. Add them back in when you need support again. Think of it this way—when you're on vacation, you likely can take a mini-vacation from your supplements. I do. On the other hand, when you're stressed out, not sleeping well, or sick, you need more supplemental support. Adding in a few essential supplements makes up for these deficiencies. Below are three that I've found very supportive:

- **Multivitamin/multimineral—without folic acid.** Your genes depend on specific nutrients in order to work properly. Many of us aren't getting enough of those nutrients for optimal genetic function. A good multivitamin/multimineral can go a long way toward solving that problem. Just be sure that you're getting one with no folic acid, which—as you know by now—can dirty your genes instead of clean them. Instead, choose a multivitamin with methylfolate and folinic acid, which are the best forms of supplemental folate available. Iron can be inflammatory, so ideally find a multi without iron (unless you know that you're iron-deficient).

—Take a multivitamin only if you feel tired, have brain fog, or otherwise sense that you need extra support. If you feel great, don't take it.

—Take one-fourth to half the suggested daily dose of your multivitamin with breakfast—if you feel you need it.

—Take another one-fourth to one-half of your multivitamin at lunch—again, only if you feel you need it.

—Never take a multivitamin within five hours of bedtime. The B vitamins can be quite stimulating and prevent you from experiencing deep sleep.

- **Electrolytes.** Many people are deficient in electrolytes, your body's electricity carriers, which are sodium, potassium, chlorine, calcium, magnesium, and phosphate. When you're low in electrolytes, your electrical energy is low. Common signs of electrolyte deficiency are muscle contraction, irregular heartbeat, mental and physical fatigue, brain fog, frequent urination, urinating within minutes of drinking water, feeling dizzy upon standing, and not sweating well. You should be taking an electrolyte supplement that contains, at a minimum, potassium, magnesium, chloride, sodium, and taurine, with no sugar, food coloring, or artificial anything. Taurine is needed as it helps carry electrolytes.

—Take the electrolytes before exercise or upon waking.

—If you don't have the symptoms listed above, and if you're not exercising or using the sauna, stop taking electrolytes.

—If you get constipated from electrolytes, you need to either drink more water or skip a day.

- **Adaptogens.** Herbal compounds that support your ability to handle stress are known as *adaptogens*. Common adaptogens are ashwagandha, rhodiola, Siberian ginseng, passionflower, and wild oats. Vitamin B_5 and vitamin C are also supportive for stress.

—Adaptogens are best taken daily, since they give you the resources to be resilient under stress. You might want to skip them while on vacation, since your stress levels should be lower then.

—Take with breakfast or, if you're quite stressed, with both breakfast and lunch.

In addition to taking the above supplements as needed, get yourself *off* as many medications as you can—but always with professional help.

- **Stop taking medications that weren't prescribed by a health-care professional.** Some over-the-counter medications, such as antacids, need to be tapered down, so work with your health-care provider to begin that process. Stopping certain medications cold turkey can create a rebound effect—that is, your symptoms come back superstrong after you stop taking the med that was designed to suppress them. This isn't fun, so avoid it by tapering off your meds gently—and *only* under a professional's care.

- **For prescription supplements and medications, ask your provider if you can stop or taper off.** Don't stop taking any prescription medication or supplement without your provider's permission and assistance. Your health could be at risk if you go cold turkey here.

Detoxification

- **Avoid plastic for use with food.** This applies to all containers that you use for cooking, storing, eating, and drinking.

- **Rely on stainless steel, glass, or clay.** Again, this applies to all kitchen uses.

- **Avoid nonstick pans or cookware.** The two-part trick is to not cook on high heat and to remove the pan from the heat for a few minutes before flipping or plating the food. It will slide right out!

- **Avoid air fresheners and scented products.** Product smells are pervasive these days—in soaps, dryer sheets, toilet paper, paper towels, and much more. Since when did clean smell "clean"? Marketing has convinced many that if something is clean, it has to have a smell. No. If something is clean, it shouldn't have any smell. If it smells, it's making your genes dirty.

- **Avoid pesticides, insecticides, and herbicides.** These are every-where. Foods. Schools. Parks. Workplaces. Start by making a change in your own backyard. The cheap combination of vinegar and water is a great weed killer, as is a propane torch. Healthy soils grow healthy plants and don't require chemicals.

- **Investigate your environment.** Look for possible sources of mold or other toxins: damp areas, actual mildew or mold patches, or water spots on floors, walls, or ceilings. Make a plan for cleanup and/or removal. Also check www.scorecard.org for the most common chemicals in your local environment. Then protect yourself from them.

- **Sweat.** Do this however you can. Exercise, fast walks with lots of warm clothes, and Epsom salt baths are all good options. If you're taking electrolytes, consider hot yoga or a sauna at 120 degrees Fahrenheit (or similar low heat) for as long as you can comfortably tolerate, placing a towel on the bench to keep the wood clean. Be sure to shower and wash with soap afterward. While you're in the sauna, focus on your breathing and try using a massage roller on your muscles or dry-brushing your skin. Never force yourself to spend time in a sauna. The moment you feel done, you *are* done. If you feel that way after thirty seconds, that's fine. Try again the next day. Rest for an hour or two afterward. Don't engage in exercise or sex during that time.

Sleep

- **Your ideal bedtime is 10:30 P.M.** If you're going to bed much later than that, start going to bed earlier in half-hour increments every other day.

- **Improve your sleep quality.** The following strategies, in combination, should make a significant difference:
  —Stop eating three hours before bedtime, unless you have a fast MAOA. If you have a fast MAOA—and if you're *not* sleeping through the night—enjoying a light snack within an hour of bedtime may be helpful. Just a few bites of that evening's leftover dinner is sufficient.

—Drink no caffeine after 2 P.M. Ideally, none at all. Period.

—Stop all electronic activity at least one hour before sleep, and put your devices on airplane mode until morning.

—Install a blue-light filter on computers, phones, and devices.

—Turn off all night-lights.

—Crack open your window for fresh air. If you're cold, get a warmer blanket.

—Block bright illumination from street lights or neighbors—ideally, ask your neighbors to turn those lights off. They're ruining your deep sleep.

—Ask someone if you snore or are a mouth-breather at night. If the answer is yes, talk with your dentist. Snoring and mouth-breathing make for a poor night's sleep and, as we saw, a dirty NOS3.

—Don't take a multivitamin before bed, as it might keep you awake at night. Supplements such as tyrosine and some herbal stimulants can also keep you awake.

—Track your sleep with the Sleep Cycle app or the ŌURA ring (see the Resources section). Tracking your sleep helps you spot trends. I've spotted many trends in my own sleep through tracking, and have altered my habits to obtain more deep sleep and REM. I've improved from averaging six minutes—yes, only six minutes!—of deep sleep and an hour of REM per night to now averaging forty-five minutes of deep sleep and three hours of REM. Good news: I've shared all my tricks with you here in this book.

Stress Relief

- **Get outside.** Go for a walk, play a sport, meet a friend, or just admire the beauty around you. In the summer, spend fifteen minutes a day out in the sun with exposed skin and no sunscreen. After that, lather up with a healthy sunscreen. (See the Resources section.)

- **Do a simple, comfortable stretching routine for five minutes each day.** The yoga sequence Sun Salutation is great, especially first thing in the morning.

- **Breathe deeply.** Focus on your breath. Breathe only through your nose at a slow and steady rate. Be aware if you're holding your breath, snoring, yawning, or mouth-breathing—and deliberately change your breathing. You should feel the air coming in slowly through your nose and then slowly going out. When people are stressed, a common response is to breathe faster and more shallowly from the chest up, rather than slowly and deeply from the belly. Work to reverse that pattern so that you continue to breathe deeply and slowly even when stressed. It's terrific stress relief and will also enable you to focus better and think more clearly.

Here's a simple five-minute exercise you can do when you're feeling stressed or anxious, have cold hands and feet, can't unwind, and/or have a dry mouth:

> While sitting in an upright position or lying down flat on your back, position one hand on your chest and one on your belly so you can feel how your hands move. Your belly hand should move first followed by your chest hand. Focus only on your breathing. Count each breath in/out as one.
>
> Notice the slightly cold air coming into your nose and the slightly warm air leaving your nose. Then start slowing down your breaths deliberately. You want to feel slightly out of breath, as though you were walking up a hill.
>
> When you're ready, slow down your breathing even a bit more. Breathe so softly that you barely feel the air entering and leaving your nose. Continue doing this until your five-minute timer goes off. During and after, you should sense warmer hands and feet, a less congested nose, increased saliva in your mouth, and an overall feeling of calm.

The next time you need a break at work or home, practice this simple breathing exercise to reset your circulation and create a calming mindset.

A Typical Soak and Scrub Day

Below is a sample schedule of a Soak and Scrub day. Use it to create your own schedule of healthy choices. I've found that what gets *scheduled*, gets *done*. Be sure to use the results from your Laundry List 1 to determine how you should best eat to support *your* genes. Use the Gene Meal Guide in the next chapter.

- **Waking up.** Wake up naturally with the sun, or set your Sleep Cycle alarm to wake you when it's optimal for your body. (The Sleep Cycle alarm tries to wake you up when you're in light sleep, but no later than your set alarm time.)

- **Morning routine.** Listen to your body as you start your day.
  —Drink 4 ounces of water with 1 teaspoon of apple cider vinegar or freshly squeezed lemon juice.
  —Do the Sun Salutation.
  —Eat breakfast, but only if you're hungry.

- **Breakfast.** If you're not hungry for your usual breakfast, skip it. Eat later.
  —Don't eat because you "have to"; eat when you start to notice that you're getting hungry. I typically wake up at 7 A.M. and eat breakfast anywhere from 10 A.M. to 11:30 A.M.
  —Some days I don't eat breakfast at all. I tune in to how I'm feeling. Clear-headed, focused, not hungry? Don't eat. Starting to get brain fog, tired, a bit hungry? Eat.
  —Don't wait to eat until you're starving or get the chills. Those symptoms mean that your blood sugar tanked, and you're likely to binge on carbs to get your sugar back up, which will create a yo-yo effect of sugar spikes and crashes that will likely last the rest of the day. Try to keep things on an even keel. The key is awareness—learning to listen to your own body. It may take a while for you to build that kind of awareness, but you'll be surprised how quickly it comes once you start routinely asking yourself, "How am I feeling now?"

- **Work.** Take a bottle of filtered water with you, augmented with electrolytes. You can use sea salt to start. (See the Resources section for options.)
  —Before starting work, go for a ten-minute walk to get some air and motion.
  —Focus on being productive. Eliminate distractions to earn yourself free time later. Identify the top three things you need to accomplish each day. Then accomplish them. If you put more than three things on your list, you may have trouble getting them all done, which can be frustrating and reinforces the idea that you can't control your day. Stick to three!
  —Say no to anything that distracts you from your main goals and agenda. Just say no. You'll be amazed at your productivity.
  —Stand up every hour and move around for a few minutes. Maybe do some push-ups or go up and down a flight or two of stairs. Even better—go outside to get some fresh air.

- **Lunch.** This is likely your largest meal of the day.
  —Don't use electronics. No driving. Eat lunch sitting down, conversing with others.
  —Chew your food well.
  —Take your time. Enjoy your meal.

- **Postwork.** Plan a nonelectronic activity for yourself when you've finished work obligations for the day.
  —Exercise, read, hike, or make progress on a hobby.
  —Do grocery shopping, laundry, or housecleaning.

- **Dinner.** Eat based on your activity for the day and how you're feeling.
  —Consult the Gene Meal Guide in chapter 13 and eat accordingly.
  —Don't eat within three hours of bedtime unless you have a fast MAOA (based on Laundry List 1), in which case have some hummus and carrots or a few bites of leftovers from dinner within an hour of bedtime.

- **Evening routine.** How you conclude your evening impacts how you'll sleep at night.
  —Filter the blue light out of your screens. Put all your electronics on

the Night Shift mode built into your phone, or install the f.lux app on all your devices.

—Write down what you're thankful for that day.

—Meditate for five minutes.

- **Bedtime.** Time going to bed so that you get between seven and eight hours of sleep. Go to sleep when you're tired. Don't fight it. (But remember, your goal is to be asleep by 10:30 P.M.!)

  —Drink a glass of filtered water.

  —Put your phone on airplane mode. Set your Sleep Cycle app. Turn off the Wi-Fi.

The preceding suggestions were targeted toward regular weekdays. Be conscious about how you schedule other days as well. Here are some suggestions:

- **Weekend.** Keep your sleep and wake times consistent with your weekday schedule.

  —Honor your weekend. Don't work unless it's absolutely essential to meet a deadline.

  —Write in your journal: What are you most thankful for in the past week?

  —Organize for the coming week. Shop. Do laundry. Clean the house and yard. Have the entire family pitch in. Assign chores and delegate routine tasks off your plate.

  —Plan an activity for each day with friends, family, yourself. It can be resting, a "staycation"—whatever is fulfilling for you.

- **Vacation.** Plan ahead. Talk out your needs and wants.

  —Where would you like to go?

  —When will the kids be out of school? Block school breaks out on your calendar, if possible. I'm self-employed, so I have some flexibility. Once I started blocking school breaks out on my calendar, our family life improved significantly! Now I always plan around my kids.

- **Spontaneous day.** Play hooky every so often. Surprise your partner and kids.

—Go skiing for the day, have a family picnic, or do a city excursion—something totally fun that says, "Screw it, I'm playing today."

Notice that the key to this schedule is *balance*. You need time to work and time to rest, play, and relax. You eat and sleep based on your body's natural rhythms, but you also help your body by setting up a routine. If you have a job that keeps you seated for long hours at a time, you honor your body's need for movement every sixty minutes or so. When you eat, you make it a time of relaxation and pleasure, so your body switches out of stress mode and into relaxation mode. Remember, stress is a real, measurable, physical factor in your health. This schedule helps you achieve true stress relief—and your genes will thank you for it.

Soak and Scrub: Week 2

Continue your routine from Week 1, with these changes/additions:

Food

- **Take steps to digest your food better.** Still have some gas and bloating with meals? Perhaps you need more digestive support. Upon waking in the morning, mix 4 ounces of filtered water with 1 teaspoon of unfiltered apple cider vinegar. Sip until you feel a slight warmth in your stomach; then stop. Do *not* do this if you have (or suspect) a stomach ulcer.

- **Focus on eating in peace, conversing with friends or family, or making the most of your alone time.** No electronics at the table, please! Enjoy mealtimes for what they are—chances to nourish you, your cells, and your genes. There are no shortcuts to good digestion. We sometimes fall for fast food, for working while eating, just shoveling that burger down. Don't go that route. Food isn't something to shush your stomach with so that you can get back to work. Enjoy the opportunity to nourish *you*. This is also a great time to interact with your colleagues or partner and your children. Kids are kids only once—ever. Mealtimes are a great way to build a healthy family—with food and conversation.

Supplements

- **Add in liposomal glutathione.** This nutrient, which many people are low in, helps countless important pathways. Don't take it unless you're already taking a multivitamin, however. The multivitamin provides the nutrients you need to help you use the glutathione.
  —Start with just 3 drops before breakfast or before lunch for three days.
  —If you feel no change or only a slight improvement, increase to ½ teaspoon a day for three days.
  —If still no change or only a slight improvement, increase to 1 teaspoon a day for three days. If you feel an improvement, continue for three weeks and then stop.
  —If you feel outstanding, stop taking glutathione until you feel that you need it again.
  —If you feel worse, stop taking it until you've added molybdenum and digestive enzymes for two weeks (see below). Then try again.

- **Take molybdenum if your breath, armpits, and/or gas smells of sulfur, or if you're sensitive to sulfites.** Start with 75 micrograms of molybdenum using The Pulse Method. This should resolve the problem within a few days. If not, continue the molybdenum, increase the amount of molybdenum taken, reduce sulfur-containing foods for a week, and then try the liposomal glutathione again.

- **Add in digestive enzymes—if needed.** If you're still experiencing gas, bloating, or belching during or after meals, you need additional digestive support. Consider taking betaine hydrochloride (HCl) along with pancreatic enzymes. If you're intolerant to fats or oils, add in lipase or about 250 milligrams of ox bile.
  —Take betaine HCl, digestive enzymes, and ox bile with meals. If you're going to eat a light meal, though, you might not need them. You'll learn from experience when you need additional support and when you don't.
  —Have an ulcer? Do *not* take betaine HCl or digestive enzymes until your stomach ulcer is healed. Use zinc carnosine, aloe vera gel, and L-glutamine to help heal the stomach ulcer.

Detoxification

- **Avoid household cleaners.** The basics work well: hot water, unscented soap, vinegar, salt, baking soda. Remember, if something smells, it's making your genes dirty.

- **Get a water filter.** Using a multistage water filter is a great and inexpensive way to drink clean water. Bottled water is usually of poor quality and is horrible for the environment, due to packaging and heavy shipping requirements. Don't use Brita-style water filters because they only filter chlorine, are comparatively more expensive than other filters, and use plastic. **Install a filter in your shower to remove all chlorine from your water.** Chlorine is harmful to your lungs, skin, and hair. Within a week of putting in the filter, you'll notice that your skin and hair feel better. You may not even need any lotions or creams once your skin gets used to a chlorine-free existence.

- **Get a HEPA filter vacuum.** A cheap vacuum blows around more dust than it picks up. Get a highly reviewed vacuum and enjoy the benefits for years. It's an investment, but it will last a long time—and because of it, so will you! Ideally, limit carpet in your home. If you can, consider replacing it with tile, stone, or wood to reduce the dust and chemicals that can irritate your genes.

- **Clean or replace your furnace air filters.** Don't skimp here. A dirty furnace puts more work on your GST/GPX gene, which then dirties all your other genes.

- **Clean your air ducts.** If you use forced air to heat your home and you haven't cleaned your ducts in two years, that job needs to get done.

- **Clean the sink traps and drain pipes under every counter.** Unscrew them, take them apart, and brush them clean with hot soapy water. Scrub inside the pipe, too. You'll be amazed how full of nasty stuff these elements are!

Sleep

- **Continue to move your bedtime toward 10:30 P.M.** You'll be amazed at how much energy you gain when you sleep during the night and wake up in the morning, rather than pushing your circadian rhythms out of balance with the cycles of the sun.

- **Rise with the sun.** Allow the sun to flood into your room every morning. Put your blinds (if needed to block out lights at night) on automatic timers. If this isn't possible, consider a sunrise alarm clock to help create morning light.

Stress Relief

- **Meditate for at least three minutes each day before bed.** See my suggestions in the Resources section of www.DrBenLynch.com for apps and other supports to help you, unless you prefer to use traditional methods of meditation. The key is to meditate consistently. Three minutes each day is more effective than twenty minutes once a week.

- **Go on a "news fast."** Stop watching the news and engaging in negative conversations. Let's face it: most news is negative. Flooding your brain with this information is hurting you. I haven't watched the news for more than ten years. I still stay connected to what's going on, and so can you. Set Google Alerts for headlines, or subscribe to your favorite online newspaper and opt for headlines. Read only what you really need to read. Ignore the rest.

- **Reduce time on social media.** Twice daily social media use is fine, but if you're on your favorite site any more often—and especially if you're checking it compulsively—I can practically guarantee that it's stressing you out more than it's relaxing you. Much more important, the time you're spending with "friends" online is pulling you away from real-life friends and family. Stress levels go up online—partly because the blue light on your computer is so stimulating, and also because you get into fights and read upsetting news and don't have anyone to process those things with. Stress levels go down after spending time with loved ones, because you feel secure and connected and plugged into the *real* world. Feeling skeptical? Try it just for the next few weeks as you complete your Soak and Scrub and move on to Spot Cleaning. You won't know how great it feels to clean out the social media gunk until it's gone!

Be Thorough

My family and I have done the Soak and Scrub for years, and I've shared it with clients and patients worldwide. I'm constantly getting emails about how this approach has helped people overcome health issues that they thought were theirs for life. I know it can help you, too.

But words on paper mean nothing. It takes action to feel better. Commit to honoring yourself and your health. Commit to restoring your health all the way down to your genetic level.

There's no rush. If you feel that this program is helping you and it's all you can handle for now, excellent. You're already way ahead. Stay on the Soak and Scrub as long as you feel you need to. Three weeks. A month. Ninety days. A year. As long as you keep improving and moving forward, there's no need to move on to Spot Cleaning.

In fact, I don't want you moving to Spot Cleaning until you feel that you've taken the Soak and Scrub as far as you can. If you're still getting results, stick with it. Don't move on to Spot Cleaning unless and until you feel you've hit a plateau.

Think of it this way. The Soak and Scrub is how I live each day, every day. It's not a temporary program. It's a lifestyle to maintain clean genes. Most of the time, I feel great and it provides what I need despite my dirty genes. When I get especially stressed, am sick or injured, face extra exposure to chemicals, or am just not feeling right, I evaluate which of my genes are dirty by tuning in to how I feel and completing Laundry List 2. From there I move to Spot Cleaning. Once I recover, I move right back into my usual daily routine of Soak and Scrub.

You wear your favorite clothes for a day or two and toss them in the washer when you change. The usual soak and scrub gets the job done, typically. Sometimes, though, you get a stain and have to spot-treat. You do so and then go back to the usual soak and scrub. This is exactly how I want you to think of this program. You are your own laundry. No one has ever taught you how to clean up your genes—until now.

Enjoy the Soak and Scrub! After two weeks, I know you'll feel much better!

13

Your Clean Genes Recipes

Look. I know you're busy. You want to make it easy to change your diet and feel better. You want a twenty-eight-day program that will provide you the answers you need to get better *now*.

Okay. But what happens in six months? A year? Two years? Ten?

It would be a disservice to you if I gave you a menu plan to follow. I don't know your schedule. I don't know where you live. I don't know your climate and food availability, nor do I know what foods you like, react to, or loathe.

A menu is handy—in some cases. If you're targeting a specific condition—excessive weight, leaky gut, autoimmune disease—a menu plan might be just what you need.

In this book, though, we're doing something radically different. I'm teaching you about how your body works, all the way down to the genetic level. You now know how the Methylation Cycle works. You know how important those "Super Seven" are—and you know how they get dirty.

So instead of offering a menu plan, I've created a number of super-healthy recipes for you, including many that I eat in my own home with my wife and boys. Each recipe is tagged with an indication of which gene(s) it supports, and which gene(s) might get a bit dirty from it.

Why would I give you a recipe that might make some of your genes a bit dirty? Well, it's nearly impossible to make every recipe support every gene. Depending on which gene or genes are giving you trouble, you'll want to focus on certain recipes and avoid others.

For example, some of the recipes call for tomatoes. If you have a dirty DAO, perhaps you should eliminate tomatoes from the recipe or find a different recipe to support your dirty DAO. If you have a clean DAO and love tomatoes—well, we have a fantastic recipe for you!

How to Determine Which Recipe to Use

- At the beginning of your Clean Genes Protocol, you'll complete Laundry List 1. Two weeks later, you'll complete Laundry List 2. Each list will help you zero in on which of your genes are dirty.

- Find the recipes that support those dirty genes. See which ones sound good to you, and add similar ones that you already use at home.

Gene Meal Guide

The following general meal tips apply to everyone, regardless of which genes are dirty:

- Fast twelve to sixteen hours a day.
- Eat only when you're calm and relaxed.
- Focus on eating. No work. No electronics. Chewing and conversation only.
- Eat a maximum of three meals a day—no snacks.
- Eat until you're 80 percent full.
- Learn to distinguish cravings from true hunger.
- Eat only when you're hungry.
- Don't eat within three hours of bedtime (unless you have a fast MAOA; then you might need a light snack an hour before bed).
- Make sure each meal has a balance of protein, carbs, and fat.
- Use organically grown foods as much as possible, or at least avoid the Dirty Dozen—the foods most contaminated by industrial chemicals according to the Environmental Working Group (www.ewg.org).

Now let's look more specifically at how you can choose recipes and plan meals that target *your* dirty gene(s). Remember, too, that each recipe in this chapter identifies the genes that it best supports.

Dirty MTHFR

- Any recipe with leafy greens or beans
- Any recipe that supports PEMT

Slow COMT, Slow MAOA

- Breakfast with balanced protein, carbs, and fat
- Lunch with balanced protein, salad, and fat
- Dinner with little protein, more salad, and fat

Fast COMT, Fast MAOA

- Breakfast with balanced protein, carbs, and fat
- Lunch with balanced protein, carbs, and fat
- Dinner with balanced protein, carbs, and fat

Dirty DAO

- Recently prepared foods only; no leftovers
- Only *fresh* seafood and meat, rinsed and dried before cooking
- Any recipe low in histamine-containing foods
- Any recipe that can be adapted to remove or reduce the amount of higher histamine foods

Dirty GST/GPX

- Any salad recipe or recipe with eggs, leafy greens, and/ or cruciferous vegetables

Dirty NOS3

- Any recipe that supports GST, MTHFR, or PEMT
- Any recipe that balances COMT and MAOA
- Any recipe that contains nuts and seeds

Dirty PEMT

- Any recipe with eggs, beets, quinoa, or lamb
- Any recipe that supports MTHFR

Clean Genes Recipes

Whenever possible, select organic ingredients and use filtered water for cooking. Conventionally "factory-farmed" meats, fish, and produce will dirty up your genes, as will unfiltered water. In addition, I recommend that you abandon standard table salt for Himalayan or Celtic sea salt, both rich in minerals. If food is to be your medicine, you must eat "clean," health-supportive food!

BREAKFAST

Tunisian Breakfast Soup with Poached Egg

This is a riff on a popular Tunisian chickpea breakfast stew, leblebi. A fried or hard-boiled egg can be substituted for poached, if you prefer. For authenticity, the hot sauce should be harissa, available in most supermarkets.

This hearty breakfast choice supports all your genes. If you have a dirty DAO, you may need to eliminate the hot sauce (unless you have a sauce that works for you).

Serves 4

4 cups chicken or vegetable broth (homemade or store-bought)

1 15-ounce can chickpeas, drained

4 cups beet greens or mustard greens, chopped into 2-inch pieces

1 tablespoon ground cumin

1 teaspoon paprika

½ teaspoon coarse sea salt, or more to taste

1 teaspoon hot sauce

Water to fill a pan 3 inches deep

2 tablespoons freshly squeezed lemon juice

4 large eggs, cold

4 thick slices gluten-free bread, toasted

1. In a small saucepan, heat the broth over medium heat. Add the chickpeas, greens, cumin, paprika, salt, and hot sauce. Cook until the greens wilt.

2. To poach the eggs, fill a pan with water 3 inches deep. Add the lemon juice, stir, and bring to a simmer over medium heat. Cooking 2 eggs at a time, crack the eggs into the water. Poach the eggs for approximately 3 minutes. (For a runny yolk cook for 2 minutes; for a set yolk, cook for 4 minutes.) Gently remove the eggs with a slotted spoon and place them on a paper towel.

3. Place the toast in 4 bowls. Distribute the chickpea soup from the saucepan over the bread. Top with an egg. Serve with additional hot sauce.

Dr. Lynch's Breakfast Smoothie

Mmm ... tangy berries, almond milk, and some protein-rich seeds for fiber and texture. Quick, easy, nourishing—the perfect breakfast if you're on the go.

This quick-and-easy smoothie supports all your genes. Add more protein powder if you need to support a fast MAOA and/or a fast COMT, and less if you have a slow COMT and/or a slow MAOA.

Serves 2

3 cups almond milk

½ cup frozen blueberries

½ cup frozen raspberries

2 tablespoons chia seeds

2 tablespoons flaxseed

2 tablespoons hemp seeds

1 to 1½ tablespoons pea protein powder

Add all ingredients to the blender and blend until smooth. Serve and enjoy!

Scrambled Eggs with Kale and Carrots

This dish offers the homey comfort of scrambled eggs with some extra fiber and nutrients from the vegetables. You'll benefit from lots of choline in the eggs and methylfolate in the kale.

This breakfast supports all genes. Those with a dirty DAO may want to skip the hot sauce.

Serves 2

2 tablespoons ghee, divided

5 to 6 eggs

½ teaspoon coarse sea salt

⅛ cup water

1 clove garlic, diced or pressed through a garlic press

½ yellow onion, thinly sliced

1 bunch kale, divided, stems cut into ¼-inch pieces and leaves cut into 1¼-inch pieces

1 large carrot, peeled and cut thinly into half moons

3 slices cooked ham or bacon, chopped

¼ teaspoon freshly ground black pepper

Hot sauce (optional)

1. Heat 1 tablespoon ghee in a pan over medium-high heat.

2. While the pan is getting hot, beat the eggs with the salt and water.

3. Pour the mixture into the hot pan and gently mix the eggs with a spatula until they're cooked through. Leave them in the pan, removed from the heat, and start working on the greens.

4. In another pan, heat the remaining ghee over medium-high heat. When the pan is hot, add the garlic and onions. Cook until lightly browned. Add the kale stems, carrots, and ham or bacon. After the carrots become slightly tender, add the kale leaves and mix in. Season

with pepper and additional salt to taste. Cover the pan and turn the burner off. Let the mixture set for another 3 to 4 minutes.

5. To serve, put the scrambled eggs in bowls and top with the vegetable mixture. If desired, add some of your favorite hot sauce.

Escarole and Feta Frittata

This flavorful frittata can be served hot, warm, or at room temperature. Add ham or sausage for added protein.

As is, this dish is good for GST/GPX, PEMT, both fast and slow COMT, and both fast and slow MAOA. If you have a dirty DAO, remove the cheese and mushrooms and make sure to use fresh (rather than cured) ham.

Serves 4

8 eggs

4 tablespoons almond milk

4 tablespoons sheep's or goat's milk feta cheese, crumbled to ½-inch pieces, divided

½ teaspoon coarse sea salt

½ teaspoon freshly ground black pepper

4 tablespoons ghee

2 tablespoons chopped onion

6 medium mushrooms, chopped into ½-inch pieces

1 pound escarole, cut into ½-inch pieces

½ cup diced cooked ham, or 2 cooked hot or sweet sausages cut into ¼-inch pieces

1. Preheat the oven to 475° F.

2. In a small bowl, beat the eggs, almond milk, half the feta, salt, and pepper.

3. In an ovenproof 12-inch skillet, heat the ghee. Add the onions and sauté over medium heat until translucent, about 5 minutes. Add the mushrooms and sauté for an additional 5 minutes. Add the escarole and cook until wilted, about 5 to 7 more minutes.

4. Add the ham or sausage and stir to combine, spreading the resulting mixture evenly in the skillet.

5. Pour the egg mixture over the vegetables and meat and cook until the eggs begin to set.

6. Sprinkle on the remaining feta. Place the skillet in the hot oven and bake for 5 minutes, until the frittata is firm but not brown.

Quinoa Porridge

A fast, tasty breakfast and a healthier hot cereal option than many traditional alternatives. For additional protein, serve it with a side of bacon, a glass of goat's milk, or an egg.

If you have a light appetite first thing in the morning, this dish is for you. While it doesn't directly support your genes, neither does it dirty them with excessive food. It's a great breakfast to help you transition from traditional breakfast cereals.

Serves 2

1¾ cups water, plus more for rinsing

1 cup quinoa

½ teaspoon coarse sea salt

1 tablespoon ghee, for garnish

Raisins, for garnish

Almond or goat's milk, for garnish

Maple syrup (optional)

1. Put the quinoa in a small pot and add a bit of water to rinse it. Dump the water and keep the rinsed quinoa in the pot.

2. Add the rest of the water and salt to the quinoa and bring it to a boil. Reduce to a simmer and then cover. Cook for 17 minutes.

3. Serve in bowls topped with ghee, raisins, and almond or goat's milk. Add a drizzle of maple syrup if desired.

Nutty Oatmeal

A quick and easy healthy start to your day made with oats and lots of nuts and seeds.

A good way to support your NOS3, slow COMT, and slow MAOA. Those with a fast MAOA or a fast COMT should add a sausage patty, bacon, or hard-boiled egg for extra protein.

Serves 4

4 cups water

1 tablespoon coconut oil

1 tablespoon ground cinnamon

1 teaspoon ground allspice

1 teaspoon ground nutmeg

¼ teaspoon ground turmeric

1 tablespoon vanilla extract

2 tablespoons almond butter

2 cups gluten-free rolled oats

¾ cup flaxseed

½ cup raw pumpkin seeds

¼ cup raw sunflower seeds

½ cup raw walnut pieces, chopped

¼ cup unsweetened coconut cream, or more to taste

½ cup chopped pistachio nuts, for garnish

¼ cup chopped almonds, for garnish

1. In a medium-size saucepan, heat the water, coconut oil, spices, vanilla extract and almond butter. Bring to a slow boil, stir, and lower the heat to simmer.

2. Add the oats, seeds, and walnuts. Cover and cook for 10 minutes, or until the mixture is the consistency you prefer. Serve with the coconut cream and garnish with the pistachios and almonds.

Bedeviled Breakfast

The smoked trout fillets that this recipe calls for can be found, packaged, in the fish section of the supermarket.

This recipe is a fantastic way to support GST/GPX, PEMT, both fast and slow COMT, and both fast and slow MAOA. This recipe is neutral for DAO, but if you're histamine-sensitive, don't include the trout, mustard, and tomatoes.

Serves 4

8 fresh eggs

Water

3 tablespoons mayonnaise

1 tablespoon Dijon mustard

¼ teaspoon hot sauce

1 teaspoon coarse sea salt

½ teaspoon freshly ground black pepper

1 teaspoon paprika, for garnish

4 ripe orange or yellow tomatoes, sliced

1 small sweet red onion, very thinly sliced (optional)

12 radishes, halved

12 ounces smoked trout, cut into 1-inch pieces

4 handfuls mixed baby greens or arugula

1. Place the eggs in a heavy-bottomed saucepan and cover them with at least 1 inch of cold water.

2. Bring the water to a full boil.

3. When there are big bubbles, remove the pot from the heat and cover. Let the pot stand untouched for 15 minutes. Remove the boiled eggs from the water. Transfer to a bowl of cold water for 10 minutes.

4. Peel the eggs and cut them in half lengthwise. Gently remove the yolks. Mash the yolks with the mayonnaise, mustard, and hot sauce. Add the salt and pepper, adjusting amounts to taste.

5. Fill the whites with the yolk mixture. Dust the eggs with paprika.

6. On a platter, arrange the eggs, tomatoes, onion, radishes, trout, and greens.

Gingered Green Smoothie

Zingy sweet creaminess is your reward with this delicious smoothie.

This nourishing breakfast supports MTHFR, GST/GPX, a slow COMT, and a slow MAOA.

Serves 1

½ cup peeled, pitted, chopped avocado

½ cup chopped fresh parsley

¼ cup chopped fresh basil

½ cup stemmed, chopped kale

½ teaspoon grated fresh ginger

1 teaspoon freshly squeezed lemon juice

½ cup almond milk

1 teaspoon MCT (medium chain triglyceride) oil

2 tablespoons pea protein powder

In a blender or food processor, blend all ingredients until smooth.

LUNCH OR DINNER

Root Vegetable Soup

The sweet potatoes, carrots, celery root, and tarragon add a lovely sweetness to this chicken soup for the soul! The Jerusalem artichoke is an amazing vegetable for liver support, and for your microbiome, too.

Enjoy this hearty soup as you support all your genes. Add chicken breast or other meat of your choice to support a fast COMT and/or a fast MAOA.

Serves 4

2 tablespoons coconut oil

1 onion, chopped

2 cloves garlic, minced

3 sweet potatoes, peeled and cut into bite-size pieces

3 carrots, peeled and cut into bite-size pieces

3 parsnips, peeled and cut into bite-size pieces

2 turnips, peeled and cut into bite-size pieces

1 celery root (celeriac), peeled and cut into bite-size pieces

3 Jerusalem artichokes, scrubbed, peeled, and cut into bite-size pieces

1 quart chicken broth (homemade or store-bought)

Water, as needed

3 tablespoons chopped fresh tarragon

2 tablespoons chopped fresh parsley

1 teaspoon chopped fresh thyme

1 teaspoon coarse sea salt, or more to taste

1 teaspoon freshly ground black pepper, or more to taste

2 boneless, skinless chicken breasts, cooked and cut into half-inch pieces (optional)

1. In a large soup pot, over medium heat, heat the coconut oil, then sauté the onion until softened.

2. Add the garlic and cook for 30 seconds. Add the vegetables and stir to combine.

3. Add the chicken broth and additional water if necessary. Add the herbs, salt, and pepper.

4. Cook over medium heat for 45 minutes or until the vegetables have softened. Add the cooked chicken breast (optional), season with additional salt and pepper to taste, and serve.

Cold Borscht

Enjoy your ultimate liquid salad—super-refreshing and yummy. It's a must at least twice in the summer season—perfect for a hot day!

This Russian version of gazpacho will nourish all your genes.

Serves 4

2½ quarts water

½ to ¾ pound cooked beets, cooled, peeled, and shredded

Juice of ½ to 1 lemon, or less if you prefer less sourness

Coarse sea salt and freshly ground black pepper to taste

1 small bunch red radishes or 6 ounces daikon radish, cut in half and then sliced thinly into half moons

1 large English cucumber, cut in half and then sliced thinly into half moons

⅓ cup fresh dill, finely chopped

⅓ cup fresh scallions or chives, finely chopped

⅓ cup fresh parsley, finely chopped

6 to 8 ounces chopped ham (optional)

1 to 2 hard-boiled eggs, chopped (optional)

Mayonnaise or plain goat's milk yogurt, for garnish
	(½ to 1 teaspoon per serving)
Additional scallions, chives, parsley, and/or dill, for garnish

1. In a large pot, combine water, shredded beets, lemon juice, salt, and pepper. Add the radishes, cucumber, and finely chopped herbs.

2. Chill the mixture in the refrigerator for at least 30 minutes, letting the flavors mix.

3. Pour the cold soup into bowls. Add the ham and eggs if desired, and garnish with mayonnaise or yogurt and additional fresh-cut herbs.

Thai Coconut Chicken Soup

An exceptional blend of vegetables and spices makes this a comforting and nourishing soup. For a variation, try serving it over cooked basmati rice.

This is a nourishing soup for all your genes, but if you have a slow MAOA or a slow COMT, cut back on the shrimp and chicken at dinnertime.

Serves 4

2 14-ounce cans coconut milk

1½ cups chicken broth (homemade or store-bought)

¼ cup green curry paste

2½ tablespoons freshly squeezed lime or lemon juice

1 tablespoon freshly grated ginger

1 pound chicken breast, sliced thinly, or 1 pound fresh shrimp, peeled

1 large carrot, cut in half and sliced into ¼-inch half moons

2 stalks celery, sliced into ¼-inch pieces

2 baby bok choy, chopped into 1-inch pieces

¼ cup fresh cilantro, chopped, for garnish

¼ cup fresh basil, chopped, for garnish

1. In a pot over medium-high heat, whisk together the coconut milk, chicken broth, green curry paste, lime or lemon juice, and ginger. Bring to a boil.

2. Add the chicken or shrimp. Cook, stirring occasionally, for 10 minutes, or until the chicken or shrimp is cooked all the way through.

3. Add the carrot and cook for 3 minutes. Add the celery and baby bok choy; then turn off the burner, cover the pot, and let the mixture sit for 3 additional minutes.

4. Divide the soup among four bowls and garnish with cilantro and basil.

Russian "Fur Coat" Salad (Shuba)

This salad is traditionally made with salted herring, but my family and I prefer a Northwest version that uses smoked wild Alaskan salmon instead.

This is one of my favorite salads. It's fantastic for those with a slow COMT or a slow MAOA, but it will support all your genes—including a dirty DAO.

Serves 4

1 pound beets, washed but not peeled

1 large or 2 medium carrots

2 medium potatoes

1 8-ounce pack cold-smoked wild Alaskan salmon, chopped into small pieces

¼ cup finely chopped red or yellow onion

1 to 2 tablespoons grapeseed or walnut oil

¼ teaspoon freshly ground black pepper

1 teaspoon dry dill or ¼ cup chopped fresh

½ cup mayonnaise (regular or egg-free)

1. Boil the vegetables for this salad the night before you want to eat it, so you don't have to wait to cool them down. Boil the beets separately from the carrots and potatoes so they'll all cook faster and the beets won't color the other vegetables. Simmer the beets, whole, in water that fully covers them, for 40 to 60 minutes; simmer the carrots and potatoes, likewise whole and fully immersed, for 20 to 40 minutes.

2. In a 5 × 8 × 3-inch glass loaf pan, combine the fish, onion, oil, pepper, and dill. Spread the mixture evenly in the dish.

3. Once cool, shred the potatoes and add for the second layer.

4. Once cool, peel and shred the carrots to create a third layer.

5. Once cool, peel and shred the beets for a fourth layer.

6. Mix the mayonnaise with a bit of water to make a thick paste. Pour the paste evenly over the top of your salad. Cover and put it in the refrigerator for 15 to 20 minutes to let the mayonnaise layer settle.

7. Add salt if desired, and serve chilled, making sure to include all four layers in each portion. We like to use a spatula for serving.

Vegetable Nut Curry

This is an almond-rich vegan stew that can be enriched with cooked chicken or pork. It's delicious paired with a leafy green or tricolor-leaf salad.

This flavorful dish offers great support for all genes, especially a slow COMT or a slow MAOA. People with a fast COMT or a fast MAOA should add more protein. This dish is well tolerated by those with a dirty DAO.

Serves 4

4 cups water

1 cauliflower, "cored" and broken into large florets

6 small sweet potatoes, peeled and cubed

3 large carrots, peeled and cut into ½-inch chunks

1 onion, chopped

¼ cup walnut oil

1 tablespoon minced garlic

2 tablespoons finely chopped fresh ginger

1 teaspoon finely chopped jalapeño pepper, seeds removed

2 tablespoons curry powder

1 teaspoon ground turmeric

½ white cabbage, sliced

2 cups almond milk

1 tablespoon almond butter

1 cup chickpeas, cooked or canned and drained

1 teaspoon coarse sea salt, or more to taste

½ teaspoon freshly ground black pepper, or more to taste

3 tablespoons chopped almonds

3 tablespoons chopped fresh parsley or cilantro, for garnish

4 teaspoons unsweetened coconut flakes (optional), for garnish

1. To a large saucepan, add the water, cauliflower, sweet potatoes, and carrots. The water should cover the vegetables by 2 inches. Bring to a boil and cook on high heat for 7 minutes, or until the potatoes can be pierced with a fork. Remove the pan from the heat. Drain the vegetables and set them aside.

2. In a 12-inch skillet, sauté the onion in walnut oil over medium heat for about 3 minutes, or until softened. Add the garlic, ginger, jalapeño, curry powder, and turmeric. Stir to combine; then cook on low heat for 2 minutes.

3. Add the cooked cauliflower, sweet potatoes, and carrots and the raw cabbage and cook the mixture gently for 5 minutes.

4. Add the almond milk, almond butter, and chickpeas and cook for 15 minutes.

5. Add more almond milk if necessary to make sure the stew is saucy.

6. Add salt and pepper to taste. To serve, sprinkle with the chopped almonds, parsley or cilantro, and coconut flakes.

Lazy Cabbage Rolls

Preparing traditional cabbage rolls takes some time. "Lazy" rolls have the same ingredients, taste just as good, and require much less time because you don't stuff the cabbage leaves and roll them up. To simplify further, you can also leave out the rice; then the dish becomes a delicious stew of ground beef and vegetables.

This lazy dish supports all genes very well and tastes fantastic. It's a favorite winter dinner in the Lynch family.

Serves 6

1 tablespoon ghee

1 white or yellow onion, chopped

1 pound ground beef

Coarse sea salt and freshly ground black pepper to taste

1 cup shredded carrots

¼ cup chopped red bell pepper (optional)

1 medium-size head white cabbage, shredded

Optional: 1 cup white rice (or 1 cup half-cooked brown rice)

For the sauce:

1½ cups water

½ cup tomato sauce or pureed fresh tomatoes

2 to 3 tablespoons mayonnaise, sour cream, or plain goat's milk yogurt (with additional for optional garnish)

1 to 2 cloves garlic, minced

1. In a large skillet over medium-high heat, warm the ghee. Add the onions and fry them until golden. Add the beef, salt, and black pepper and let the mixture cook together for 10 minutes. If needed, drain off the fat.

2. Add the carrots and bell pepper and cook for 2 minutes. Add the cabbage and rice (optional) and lower the heat, simmering until the vegetables are tender and the rice is cooked.

3. In a bowl, combine all the ingredients for the sauce. Pour the sauce into the pan with your cooked ingredients. Bring that mixture to a boil, covered, and then adjust to a simmer for 15 minutes. If the sauce gets thicker than you like, add more water as everything cooks.

4. Divide among six bowls and serve. Add an additional dollop of mayonnaise, sour cream, or yogurt for garnish, as desired.

Vibrant Green Vegetable Salad with Lynch Family Dressing

In composing this vibrant green salad, you can use all the ingredients listed or choose just a few—or add your own favorites!

This salad supports MTHFR, slow COMT, slow MAOA, and GST/GPX. Those with a dirty DAO may need to eliminate the olives. Those with a fast MAOA and a fast COMT should add sliced chicken.

Serves 4

2 cups stemmed and ribbon-sliced kale

4 cups mixed baby greens

1 cup watercress, stemmed

2 cups arugula

1 tablespoon plus 2 teaspoons chopped fresh tarragon, divided

Lynch Family Dressing (page 263)

16 endive leaves

16 very thin asparagus spears

2 cups sugar snap peas

2 avocados, peeled, pitted, and sliced

½ cucumber, peeled, seeded, and sliced

½ cup coarsely chopped green pepper

1 large fennel bulb, thinly sliced

½ cup pitted green olives, or more to taste

1. In a bowl, combine the kale, baby greens, watercress, and arugula.

2. In a separate bowl, add 2 teaspoons of tarragon to the Lynch Family Dressing. Toss the greens with 2 tablespoons of dressing.

3. Divide the greens among four plates. Arrange the endive in four directions, tucking a bit of each leaf under the mounded greens but leaving the points out. Place the asparagus spears on the endive with the cut ends tucked under the greens.

4. In another bowl, combine the snap peas, avocados, cucumbers, green peppers, fennel, and olives with 2 tablespoons of the dressing. Spoon onto the greens. Sprinkle with the remaining tarragon.

Warm Artichoke, Asparagus, and Pine Nut Salad

A warm salad is good any time of year—especially this salad, with its array of flavors and textures.

This would make a wonderful lunch or dinner for those with a dirty MTHFR, a slow COMT, a slow MAOA, or a dirty GST. If you have a dirty DAO, you may need to reduce the amount of Dijon mustard and pine nuts, but the quantities are small so you may tolerate them just fine.

Serves 4

4 medium artichokes

Water

1 teaspoon freshly squeezed lemon juice

16 asparagus spears, bottom quarter removed

2 cups cooked wild rice

1 tablespoon ghee

2 tablespoons pine nuts

1 pound baby bok choy, sliced

For the dipping sauce:

6 tablespoons freshly squeezed lemon juice

2 teaspoons finely grated lemon zest

1 teaspoon Dijon mustard

Coarse sea salt and freshly ground black pepper to taste

½ cup olive oil

3 teaspoons flaxseed oil

1. With scissors, cut off the thorns of each artichoke and the end of the stem, leaving about one inch of the stem. Place a steamer basket in a large pot with a lid. Fill the pot with water until it reaches the bottom of the basket; add 1 teaspoon lemon juice. Place the artichokes in the steamer basket, stem end down. Put a lid on the pot and bring the water to a boil. Lower the heat to medium and steam the artichokes for 40 minutes, or until the stem can be pierced with a fork. Drain and set aside. (Keep the pot close; you'll be using it again.)

2. In the same pot, boil the asparagus in salted water until it's crisp and tender. Drain and set aside.

3. In the pot you cooked the artichokes in and then the asparagus in, combine the rice and the ghee to warm them. Once warm, add the pine nuts to the pot and combine.

4. To make the dipping sauce, in a small lidded jar, combine the lemon juice, lemon zest, mustard, salt, and pepper. Add the two oils and vigorously shake. Add more salt and pepper to taste.

5. In the middle of each of four plates, make a bed of the bok choy, spoon the rice in the center, top with the artichoke, and surround with the asparagus. Season with additional salt and pepper to taste, and serve with the lemon dipping sauce on the side.

Seared Scallops

Not all scallops are the same! Buy "dry" scallops, not "wet"—terms that refer to how the scallops were packed after shucking. Dry scallops don't have the chemical preservatives of wet, and can be identified by their pearly or pink look. Pair with basmati rice and a side dish of green beans sautéed with sliced carrots and walnuts.

This nutritious meal supports a fast COMT, a fast MAOA, a dirty NOS3, and a dirty PEMT. Those with a slow COMT and a slow MAOA should do well with this dish if they eat fewer scallops. Those with a dirty DAO should tolerate this dish well if the scallops are fresh.

Serves 4

1¼ pounds large dry sea scallops

½ teaspoon coarse sea salt

½ teaspoon freshly ground black pepper

2 tablespoons avocado oil or ghee

2 tablespoons freshly squeezed lemon juice

1 teaspoon capers (packed in salt), rinsed

1 tablespoon chopped fresh parsley

1. Wash and dry the scallops. Season them with the salt and pepper.

2. Heat the oil or ghee in a heavy 10-inch skillet over high heat.

3. Quickly place the scallops in the skillet, in one layer, not touching. Sear the scallops for 1 to 2 minutes on each side, until a golden crust forms. Remove them from the pan and arrange them on plates.

4. Add the lemon juice, capers, and parsley to the oil or ghee remaining in the pan. Cook until hot. Pour the mixture over the scallops and serve.

Picadillo

This classic sweet-and-sour Cuban dish can be made with ground pork or beef. Serve with brown rice and organic corn tortillas.

This tangy dish supports those with a fast COMT or a fast MAOA. Those with a slow MAOA and a slow COMT should eat a smaller serving for dinner, to limit their late-day protein. If you have a dirty DAO, this dish should be well tolerated because the tomatoes and olives are cooked. If you're very sensitive, though, you can skip those two ingredients.

Serves 4

4 tablespoons coconut oil

1 onion, finely chopped

3 cloves garlic, finely chopped

1 pound ground lean pork

1½ teaspoons ground cumin

1½ teaspoons ground allspice

1 teaspoon dried oregano

1¼ teaspoons ground cinnamon

1 teaspoon coarse sea salt

¼ teaspoon freshly ground black pepper

1 28-ounce can chopped tomatoes, *not* drained

3 tablespoons freshly squeezed lemon juice

2 tablespoons honey

¾ cup raisins

2 teaspoons capers, packed in salt, rinsed

2 tablespoons chopped pimento-stuffed green olives

1. Heat the oil in a medium skillet over medium heat. Add the onions and cook until soft but not brown. Add the garlic and cook for 30 seconds.

2. In a bowl, combine the pork, cumin, allspice, oregano, cinnamon, salt, and pepper, using a spoon to break up any clumps.

3. Add the pork mixture to the onion and garlic in the skillet and cook for 6 minutes.

4. Add the tomatoes, lemon juice, honey, raisins, capers, and olives. Cook for 15 minutes, or until the sauce has thickened. Taste for seasoning.

Fish Stew

This is an easy stew that can be enriched with shellfish.

This yummy stew supports those with a fast COMT, a fast MAOA, and a dirty GST/GPX, NOS3, or PEMT. Those with a slow COMT or a slow MAOA should consider less protein—perhaps using only half the recommended amounts of fish and optional shellfish. Those with a dirty DAO should tolerate this dish if both the fish and shellfish are fresh and well rinsed before cooking.

Serves 4

2 tablespoons coconut oil

1 cup chopped onion

4 cloves garlic, minced

1 cup chopped fennel stalk (save the greens for garnish)

1 cup chopped carrots

2 teaspoons coarse sea salt, or more to taste

1 teaspoon freshly ground black pepper, or more to taste

2 cups fish stock or bottled clam juice

2 cups water

1 28-ounce can crushed tomatoes

5 star anise seeds

24 mussels (optional)

1½ pounds fresh cod or haddock, cut in 2-inch pieces

2 tablespoons chopped fresh parsley

12 medium "dry" (see page 254) scallops (optional)

1 tablespoon fennel greens, for garnish

1. In a 10-inch Dutch oven, heat the oil and sauté the onion until it's soft and golden. Add the garlic and cook for 30 seconds. Don't let the garlic brown.

2. Add the fennel and carrots, salt and pepper, and cook for 5 minutes. Add the fish stock or clam juice, water, tomatoes, and star anise and cook for 15 more minutes, or until the carrots are tender.

3. If using the optional mussels, add them to the pot and continue cooking until they open.

4. Remove the cooked mussels from the pot and set them aside.

5. Add the fish and the parsley. Bring the soup back to a simmer and cook until the fish easily flakes apart, about 5 minutes.

6. Add the optional scallops and cook until they turn opaque. Remove the scallops from the pot.

7. Taste the soup for additional salt and pepper.

8. When you're ready to serve, place 3 scallops and 6 mussels in each bowl. Fill the bowls with the hot fish soup and garnish with the fennel greens.

Roasted Miso Chicken and Vegetables

Asian influence on a traditional roast chicken flavors this delicious dish. Serve with brown rice or quinoa.

This flavorful dish supports those with a fast COMT, a fast MAOA, or a dirty PEMT. Those with a slow MAOA or a slow COMT should eat less chicken and more vegetables when this is served at dinner. Miso can dirty the DAO, so if yours is already very dirty, consider skipping this ingredient; however, since the miso is cooked and only a small amount is used, it should be well tolerated by most.

Serves 4

4 tablespoons white or yellow miso

½ cup sunflower or safflower oil

¼ cup honey

2 tablespoons freshly squeezed lemon juice

1 teaspoon finely chopped fresh ginger

1 teaspoon coarse sea salt

½ teaspoon freshly ground black pepper

4 chicken breasts or 8 chicken thighs, bone-in, skin-on

3 carrots, cut into ½-inch pieces

1 cauliflower, "cored" and cut into ½-inch pieces

2 teaspoons toasted white sesame seeds, for garnish

1. Preheat the oven to 425°F.

2. Line 2 baking sheets with parchment paper brushed with oil.

3. In a bowl, combine the miso, oil, honey, lemon juice, ginger, salt and pepper. Set aside 2 tablespoons of this mixture and divide the remainder into 2 large bowls.

4. In one bowl, rub the miso into the chicken and let it marinate for 30 minutes or more. Toss the carrots and cauliflower into the other bowl just before cooking. Place the chicken in a single layer on one baking sheet and transfer the vegetables to the other.

5. Roast in the oven for 30 minutes, or until the skin is crisp and the internal temperature of the chicken is 160 to 165°F. The vegetables should be tender but crisp.

6. Divide both the chicken and the vegetables among four plates. Garnish with sesame seeds.

Salmon with Ginger Vinaigrette

This is a slightly Asian take on a delicious fish dish. Accompany with basmati or wild rice and sautéed asparagus for a real feast

This dish supports a dirty MTHFR, a fast COMT, and a fast MAOA. Those with a dirty PEMT should replace the coconut oil. Those with a dirty DAO should tolerate fresh and washed salmon. Those with a slow COMT and/or a slow MAOA should eat a smaller portion of the salmon and eat more of the vegetables.

Serves 4

4 wild salmon fillets, about 7 ounces each

3 teaspoons grated fresh ginger

1 tablespoon gluten-free soy sauce

1 teaspoon sesame oil

1 tablespoon olive oil

Coarse sea salt and freshly ground black pepper to taste

2 teaspoons coconut oil

1. Preheat the oven to 450°F.

2. Wash and dry the salmon.

3. In a food processor, combine the ginger, soy sauce, sesame oil, and olive oil. Process until smooth and set aside.

4. Place a small, heavy-bottomed ovenproof skillet or cast-iron pan over high heat.

5. Salt and pepper the fish on both sides.

6. When the pan is very hot, add the coconut oil and place the salmon in the pan, flesh side down. Cook on high heat until opacity creeps one-third of the way up the side of the fish, about 3 minutes. Don't turn the fish.

7. Put the pan in the oven and bake the salmon for 5 to 7 minutes, until the flesh is opaque and firm. Using a long spatula, turn the fish onto a plate.

8. Serve with the ginger vinaigrette on the side.

Lamb Chops with Herb Sauce

The zesty, minty sauce, paired here with lamb, may also be served with pork or chicken. Roasted potatoes and sautéed green beans would pair well with this dish.

This dish supports a fast COMT, a fast MAOA, and a dirty PEMT. Those with a slow COMT and/or a slow MAOA should eat less of the lamb and more of the vegetables. Those with a dirty DAO should leave the anchovy out of the recipe.

Serves 4

5 cloves garlic, minced, divided

½ teaspoon chopped fresh rosemary

2 teaspoons coarse sea salt

½ teaspoon freshly ground black pepper

8 loin lamb chops, each about 1¼-inch thick

For the sauce:

1 cup fresh mint, chopped

¼ cup fresh cilantro, chopped

½ cup fresh parsley, chopped

1 teaspoon chopped jalapeño pepper, seeds removed

1 anchovy fillet (optional)

1 tablespoon honey

1 teaspoon freshly squeezed lemon juice

½ teaspoon hot sauce (optional)

½ cup olive oil

1. Preheat the broiler or grill to medium heat.

2. Mash two cloves of the garlic, the rosemary, and the salt and pepper into a paste.

3. Rub the mixture into the lamb chops and set aside for 10 to 15 minutes.

4. In a food processor, pulse the mint, cilantro, parsley, remaining garlic, jalapeño pepper, and optional anchovy until well combined. Add the honey, lemon juice, and optional hot sauce, and pulse briefly. With the machine going, slowly add the olive oil until incorporated. Add additional salt, pepper, or hot sauce to taste.

5. Broil the chops for 4 minutes a side for medium (3 minutes for medium-rare). The chops should be 4 to 5 inches from the flame or broiler.

6. Serve the sauce on the side.

Vegan Rice Bowl

This Mexican-style rice bowl can include cooked chicken or meat for additional protein.

This dish supports a dirty MTHFR, a slow COMT, and a slow MAOA. Those with a fast COMT or a fast MAOA should add some chicken or additional beans. Those with a dirty DAO shouldn't use tomatoes and should not use lime juice in the dressing.

Serves 4

3 tablespoons coconut oil, divided

1 tablespoon chopped onion

1¼ teaspoons minced garlic, divided

1 cup uncooked brown rice

2¼ cups water, divided

½ teaspoon ground cumin

¾ cup Lynch Family Dressing (see page 263)

1 teaspoon chopped cilantro (plus 2 teaspoons for optional garnish)

2 cups canned black beans, rinsed and drained

1 teaspoon coarse sea salt, divided

½ teaspoon freshly ground black pepper, divided

3 cups tightly packed chopped chard, kale, or escarole

3 large ripe orange or yellow tomatoes, diced

2 avocados, peeled, pitted, and sliced

1. In a medium saucepan, add 1 tablespoon of the oil and sauté the onion until soft. Add ¼ teaspoon of the garlic and cook for 30 seconds.

2. On low heat, add the rice and sauté, stirring until it turns opaque. Add the water and cumin and cook, covered, for about 30 minutes, or until the liquid has been absorbed.

3. While the rice is cooking, prepare the Lynch Family Dressing and adding 1 teaspoon chopped cilantro.

4. In a sauté pan over low heat, warm the rest of the garlic in the remaining oil. Do not brown.

5. In a bowl, combine half of that garlic oil with the drained beans. Add ½ teaspoon salt and ¼ teaspoon pepper. Set aside.

6. Wash the greens, shake off excess water, and sauté the damp greens in the remaining garlic oil, still on low heat, for 5 minutes, or until they're wilted. Season with remaining salt and pepper.

7. To assemble the dish, spoon rice into four large soup bowls. Top with consecutively placed sautéed greens, tomatoes, avocados, and black beans. Pour the dressing over the vegetables. Garnish with the remaining chopped cilantro (optional).

BASIC RECIPES

Lynch Family Dressing

¼ cup walnut oil, grapeseed oil, or sunflower oil

1 to 2 tablespoons maple syrup

1 to 2 tablespoons apple cider vinegar or tamari

2 tablespoons freshly squeezed lemon or lime juice

1 to 2 teaspoons minced garlic

1 to 2 teaspoons thinly grated fresh ginger

¼ teaspoon freshly ground black pepper

⅛ cup water

Put all ingredients in a small glass jar or bottle and shake it well. You can store this dressing in the refrigerator for weeks.

Vegetable Preparations for Entrée Sides

Most entrées in the preceding "Lunch or Dinner" recipe section mention vegetable pairings. They are suggestions only. Please follow your taste and substitute your own choices.

Enhance sautéed and roasted vegetables with herbs and spices—for example, cucumber and fennel with tarragon, carrots with cumin and cinnamon, cauliflower with curry, zucchini with basil.

The fastest vegetable preparations are wilting and sautéing. Roasted vegetables take longer, but the cooking time can be reduced by parboiling hard vegetables before seasoning and roasting.

Composed salads also make easy sides. Leftover roasted vegetables are a savory addition to a green salad. Unusual vegetables such as jicama, parsnips, shredded brussels sprouts, beets, and Jerusalem artichokes make interesting ingredients in a salad. Rice, potatoes, and grains such as millet, quinoa, and amaranth are textural enhancements. The simple "green salad" can be varied in color, shape, texture, and taste. Try including radicchio, endive, Boston lettuce, Bibb lettuce, red and white cabbage, arugula, spinach, watercress, mustard greens, and/or a mixture of baby lettuce. Dried fruit and nuts and grated cheese add sweetness and texture.

Roasted Vegetables

Potatoes, cauliflower, carrots, onions, brussels sprouts, asparagus, squash, garlic, and all root vegetables are ideal for roasting. Leftover roasted vegetables, at room temperature, make a delicious salad ingredient or snack.

1. Preheat the oven to 450° F.

2. Cut the vegetables into uniform sizes and shapes. For example, potatoes should be halved, onions should be quartered, squash and carrots should be cut into 1-inch chunks. Toss with oil; baking times will vary.

3. To save time, hard vegetables such as cauliflower, potatoes, carrots, and squash can be parboiled until just tender.

Sautéed Vegetables

Sautéing is a quick method of cooking vegetables, taking typically between 3 and 7 minutes. The key is to cut the vegetables into equivalent bite-size pieces. This ensures that they will finish cooking at the same time. The cooking time will vary depending on the type of vegetable. Green beans, zucchini, mushrooms, asparagus, corn kernels, and tomatoes will cook in a short time. Brussels sprouts, broccoli, and cauliflower cook in a medium amount of time. Very dense vegetables, such as potatoes and carrots, can be steamed or cooked in water first to make them ready for the sauté pan. If you're making a medley of vegetables, add those that have the longest cooking time first. The aim is to achieve the desired firmness and integrity of each vegetable.

1. Cut the vegetables into rounds, sticks, or bite-size pieces. They should be uniform in size.

2. In a wide sauté pan, add avocado oil. Heat the pan to medium-high.

3. When the oil starts to shimmer, add minced garlic, immediately followed by the vegetables. Don't overcrowd the pan. If necessary, cook in two batches. Toss the vegetables repeatedly and cook until fork-tender. The time will depend on the type of vegetables. One minute before removing from the heat, add herbs and spices, including salt and pepper.

Wilted Vegetables

For all tender leafy greens such as Swiss chard, spinach, escarole, baby kale, dandelion greens, mustard greens, and broccoli rabe, this cooking method works well. Allow 2 cups tightly packed greens per person. (The following recipe can be scaled up as needed.)

Serves 2

4 cups packed greens

3 small garlic cloves, thinly sliced

2 tablespoons ghee or avocado oil

1 tablespoon freshly squeezed lemon juice

Coarse sea salt and freshly ground pepper to taste

1. Remove tough stems from the greens. Wash the leaves and drain them in a colander. Don't dry them.

2. On low heat, lightly sauté the garlic in the oil. Don't allow it to brown. Turn the heat up to medium and add the greens, tossing them in the oil until they're fully wilted.

3. Add the lemon juice, then salt and pepper to taste.

14

Laundry List 2: Which Genes Need More Cleaning?

Wow, here we are again—time to run through our second Laundry List! Now that you've given your genes two weeks of great diet and lifestyle, let's dig even deeper to find out which genes might need some additional support.

Laundry List 1 was a great way to quickly evaluate which genes are dirty. It allowed you to determine which recipes to use. The goal of Laundry List 2 is to really dial in and determine which additional lifestyle, dietary, and environmental changes you need to implement, along with supportive supplementation. It isn't possible to act on Laundry List 2 without having completed Laundry List 1 and spent at least two weeks on your Soak and Scrub.

Fill out this questionnaire—again, be absolutely honest—and calculate a separate score for each gene. Use those scores to determine which genes need extra attention as you turn to chapter 15 and find out how to Spot Clean specific dirty genes.

Check each box if the condition has occurred frequently within the last sixty days or is generally true:

MTHFR

- ☐ I get shortness of breath or become red in the face after exercising.
- ☐ At times, I get exercise-induced asthma.
- ☐ My moods often fluctuate between irritability and depression.
- ☐ I can't easily tolerate alcohol of any type.
- ☐ I feel generally tired and "toxic."
- ☐ I don't eat leafy green vegetables every day.
- ☐ I tend to be able to focus and concentrate quite well—when I'm not mad or sad.
- ☐ I have difficulty falling asleep at times.
- ☐ I've had laughing gas (nitrous oxide) at the dentist's or the doctor's office and it made me feel horrible.
- ☐ When I get irritated, it takes me quite some time to calm down.
- ☐ There are some days I push it and take risks, but that's usually not my style.

DAO

- ☐ I'm often irritable, hot, or itchy after eating.
- ☐ I can't tolerate yogurt, kefir, chocolate, alcohol, citrus, fish, wine (especially red), or cheese.
- ☐ I get random joint pains that move around and come and go.
- ☐ I have skin issues such as eczema, urticarial (hives), or psoriasis.
- ☐ If I scratch my skin, I get red streaks.
- ☐ I can't tolerate many probiotics.
- ☐ I have SIBO.
- ☐ I have a lot of food allergies or food intolerances.
- ☐ I have ringing in my ears at times, especially after eating.
- ☐ I've been told I have leaky gut syndrome, Crohn's disease, or ulcerative colitis.
- ☐ I get migraines or other headaches often.
- ☐ I have a runny nose often, as well as nosebleeds.
- ☐ I can't fall asleep for several hours after eating or drinking.
- ☐ I have asthma or exercise-induced asthma.

COMT (slow)

- ☐ I feel *more* irritable after eating a high-protein diet (GAPS, Paleo).
- ☐ I'm easily irritated, and it takes me a long time to calm down.
- ☐ I routinely have (or used to have) PMS.
- ☐ I'm a very happy, enthusiastic person—but it's easy to irritate me.
- ☐ I'm not very patient.
- ☐ I've always been able to focus and study for long hours.
- ☐ I've struggled with falling asleep since I was a child. I know the ceiling patterns well.
- ☐ My doctor put me on birth-control pills to control acne or heavy bleeding.
- ☐ I have (or have had) uterine fibroids.
- ☐ Caffeine does wake me up, but I have to be careful not to drink too much or I get irritable.
- ☐ I don't like taking risks. I'm pretty cautious.

COMT (fast)

- ☐ I have difficulty paying attention.
- ☐ I get depressed quite often.
- ☐ When I get stressed out, I can calm down quickly.
- ☐ I tend to be calm most the time, but I don't always like to be.
- ☐ I'm a risk taker. I enjoy pulling stunts, because I feel awesome afterward.
- ☐ I'm the class clown. I love it when I make people laugh.
- ☐ I find myself fidgeting and moving constantly.
- ☐ I sometimes pinch myself so hard that it hurts.
- ☐ I have a hard time getting going in the morning.
- ☐ I find that I can easily get addicted to things or activities: video games, social media, smoking, drinking, shopping, drugs, gambling.
- ☐ I'm not very interested in sex.
- ☐ When it's bedtime, my head hits the pillow and I'm out like a light.
- ☐ Caffeine helps me focus and pay attention.
- ☐ I crave high-fat, high-sugar foods, and they do make me feel better—for a bit.

MAOA (slow)

- ☐ I tend to be rather aggressive.
- ☐ It takes me a while to slow down.
- ☐ I can focus for a long time.
- ☐ When I drink alcohol, I become an angry drunk.
- ☐ I'm not drawn to carbs, and I feel less irritable when I don't eat many of them.
- ☐ I'm more irritable and angry when I eat cheese and/or chocolate or drink wine.
- ☐ It takes me a while to fall asleep.
- ☐ When I do fall asleep, I tend to stay asleep through the night.
- ☐ My doctor put me on an SSRI for depression, and I got very irritable from it.
- ☐ Melatonin doesn't work well for me. It makes me feel more awake and irritable.
- ☐ Caffeine tends to make me irritable.
- ☐ Lithium helps my mood.
- ☐ 5-HTP makes me feel anxious and irritable.
- ☐ Inositol overstimulates me.
- ☐ I'm self-confident.
- ☐ I'm a man.

MAOA (fast)

- ☐ Since I was a kid, I've had a very hard time focusing and paying attention.
- ☐ I crave cheese, wine, and chocolate, and I feel better after I consume them.
- ☐ I crave carbohydrates, and they make me feel less depressed.
- ☐ I fall asleep quite well, but I tend not to be able to sleep through the night. I need a snack to fall back asleep.
- ☐ I have an autoimmune disease, such as Graves' disease, Hashimoto's thyroiditis, multiple sclerosis, or active celiac.
- ☐ I'm chronically inflamed.
- ☐ Winters and prolonged darkness affect my mood. I've been told I have seasonal affective disorder.

- ☐ I love exercising. It helps my mood.
- ☐ I'm a woman.
- ☐ I'm a worrier.
- ☐ I tend to be depressed and anxious.
- ☐ I get a bit obsessive about things.
- ☐ I have fibromyalgia, constipation, or irritable bowel syndrome.
- ☐ Melatonin works quite well to help me sleep.
- ☐ Inositol improves my mood.
- ☐ 5-HTP improves my mood.
- ☐ Lithium makes me feel more depressed.
- ☐ My doctor put me on an SSRI, and it did help.

GST/GPX

- ☐ I'm sensitive to chemicals and smells.
- ☐ I feel way better after taking a sauna or sweating heavily.
- ☐ It's easy for me to gain weight even though I eat right.
- ☐ Cancer runs in my family.
- ☐ I notice gray or white hairs coming in when I get stressed.
- ☐ I have early graying of my hair.
- ☐ I have high blood pressure.
- ☐ I just got done fighting an infection.
- ☐ I tend to be chronically stressed out.
- ☐ I have an autoimmune disease.
- ☐ I have chronic inflammation.
- ☐ I have asthma or difficulty breathing. I often feel like I can't get enough air.
- ☐ I generally feel tired and "toxic."
- ☐ I have high blood pressure.

NOS3

- ☐ I have high blood pressure.
- ☐ I've had a heart attack.
- ☐ I have diabetes, type 1 or 2.
- ☐ I have cold hands and feet.

☐ I struggle with asthma.

☐ I snore, breathe through my mouth, or have sleep apnea.

☐ I'm noticing that my memory is getting worse.

☐ I had preeclampsia during pregnancy.

☐ I have atherosclerosis.

☐ I'm postmenopausal.

☐ My moods are all over the place.

☐ I don't exercise or move much.

☐ I have an autoimmune disease.

☐ I'm chronically inflamed.

PEMT

☐ I'm postmenopausal.

☐ My estrogen levels are low.

☐ I have gallstones.

☐ I don't eat leafy green vegetables often.

☐ I don't eat eggs or meat much.

☐ I've been told I have fatty liver.

☐ I have SIBO.

☐ I'm a vegetarian or vegan.

☐ I had my gallbladder removed.

☐ I've had general pain everywhere—inside and out—for years.

☐ I don't tolerate fatty foods well.

☐ My symptoms started partway through pregnancy and have gotten worse since.

☐ My child has a congenital birth defect.

☐ Breastfeeding wore me out physically and mentally.

Scoring

Score each gene separately.

- 0 points: Excellent! This gene is amazingly clean!

- 1–4 points: This gene needs a bit of attention, but the problems are probably related to several genes rather than this particular one.

- 5–7 points: This gene seems to be a bit dirty. Paying some direct attention to this gene will likely produce results. Looking at how other genes influence this gene is important.

- 8 points and above: This particular gene is definitely dirty. Spend some time identifying all the factors that are affecting its function. Identify other genes that scored high, because they are dirtying this gene as well.

My Score

MTHFR _____	MAOA (fast) _____
DAO _____	GST/GPX _____
COMT (slow) _____	NOS3 _____
COMT (fast) _____	PEMT _____
MAOA (slow) _____	

Being Happy with Your Haplotype

Harriet, Eduardo, and Larissa, whom we met back in chapter 3, each identified one key dirty gene. But sometimes we can identify *combinations* of dirty genes—what scientists call *haplotypes*. For example:

- SNPs in both MTHFR and NOS3 give you an increased risk of cardiovascular issues and migraines, which you can address through diet, exercise, and stress relief. On the flip side, both of these genes are thrifty in the ways they conserve nutrients: with this haplotype, you typically have more folate for DNA repair and more arginine for muscle tone and infection fighting.

- The cardiovascular issues and migraines mentioned above become even more intense if you have SNPs in MTHFR, NOS3, and COMT; and your risks increase further with SNPs in MTHFR, NOS3, COMT, and GPX/GST. You don't need to panic if you have that haplotype—but you do need to make extra sure to follow the Clean Genes Protocol and give your dirty genes all the support they need. Once again, the good news is that you're conserving folate and arginine. In addition, your brain chemicals stick around longer with this haplotype, giving you much greater attention and focus.

- SNPS in both MTHFR and DAO increase histamine intolerance, putting you at risk of chronic or exercise-induced asthma. With this haplotype, you'll want to take extra care to avoid histamines in your food and your environment, and to choose the kinds of aerobic exercise that increase your lung capacity.

- The haplotype of SNPs in MTHFR, DAO, COMT (slow), and MAOA (slow) further increases histamine intolerance, putting you at even greater risk of chronic or exercise-induced asthma. Increased irritability and anxiety are also potentially present, and you need to take even greater care with diet and exercise. However, your ability to focus is incredible, and people wonder how you can concentrate for so long.

- SNPs in both MTHFR and COMT (slow) increase aggression, irritability, and your risk of estrogen-related cancers, which means you might want to incorporate extra stress relief into your life. Vacations are a fabulous way to combat the downside of this haplotype. The good news is that you're extremely productive and get things *done*. Did I say your skin looks fabulous, too?

- The haplotype of SNPs in MTHFR, COMT (slow), and GST/GPX make you still more likely to be aggressive and irritable and put you at risk for estrogen-related cancers, neurological disorders such as Parkinson's disease or multiple sclerosis, and cardiovascular disorders such as heart attack and hypertension. You can overcome these risks with the Clean Genes Protocol, but you'll want to make extra time for stress relief. On the plus side, your creativity and focus are *solid*.

- The haplotype of SNPs in MTHFR, COMT (slow), MAOA (slow), and GST/GPX further increases your risk of irritability, neurological disorders, and insomnia. When you're on, though, you're on. The things you come up with and accomplish are unreal. Some say you're a genius!

- SNPs in both MTHFR and PEMT increase your risk of pregnancy complications, gallbladder issues, SIBO, and fatty liver. These risks are increased still further if you have the triple combination of MTHFR, PEMT, and GST/GPX. With this haplotype, you have a great reason to eat more meat and eggs!

- SNPs in MTHFR, PEMT, and NOS3, in combination, further increase the risk of pregnancy complications, liver problems, and cardiovascular issues. The risk increases a bit more with the combination of MTHFR, PEMT, NOS3, and GST/GPX. The good news is that you don't have to worry about the risks even a little if you follow the Clean Genes Protocol.

- The haplotype of SNPs in COMT (fast) and MAOA (fast) increases your risk for ADD/ADHD, lack of drive, and depression. On the upside, your friends say you're the most chill and easygoing person they know.

Understanding these and many other combinations gives you the power to live a healthier, happier life. Once again, the Clean Genes Protocol allows you to make the most of your upside while minimizing the risks to your downside.

Spot Cleaning for Life

Your body is constantly changing—and your environment is too. Maybe you've had a stressful two months at work and are now passing into a peaceful time. Or perhaps you've had a quiet, pleasant summer and are now gearing up for a challenging fall. Maybe your tastes in food have changed, or perhaps you've noticed significant differences in your health thanks to your time on the Clean Genes Protocol.

Whatever your situation, your health is a lifelong journey. Don't just complete the questionnaire in this chapter, do your Spot Cleaning, and forget about it. I encourage you to return to Laundry List 2 every three to six months—as you continue to live according to the Clean Genes Protocol—and to use this questionnaire as your guide for whatever Spot Cleaning you need *throughout your life*. That's how my family and I do it—and that's how I encourage my clients to do it also. Your Laundry List 2, your Soak and Scrub, and your Spot Cleaning procedure—detailed in the next chapter—are your best tools for keeping your genes clean now and for the rest of your life.

15

Spot Cleaning:
Your Second Two Weeks

So you've completed the Soak and Scrub and you're raring to dig deeper and get more specific, all the while maintaining the gene-friendly diet and lifestyle you've begun. Terrific!

But before you move on to Spot Cleaning, I have to ask: Did you do everything—I mean *everything*—in the Soak and Scrub? If you did, you're ready to move on to Spot Cleaning while continuing to follow the Soak and Scrub principles. If you didn't, your results for Spot Cleaning will be—well—spotty.

In order for you to Spot Clean a particular gene, *all* the genes have to be fairly clean, just as you can't target one specific spot on your jeans until the whole garment is generally clean. For general cleanliness to happen in your body, you need to *faithfully* follow the Soak and Scrub approach.

Remember, genes work together—in groups and clusters. So if you decide to skip straight to this chapter, you might find yourself frustrated.

Plan of Attack

Here are important points to keep in mind as you Spot Clean:

- The cleaner your genes are, the faster you'll be able to lower the dose of your supplements, or stop them altogether.

- The dirtier your genes are, the more you'll need to start with a low dose of a nutrient and then work up to what works best for you. You might find that you need the nutrient at a higher amount for some time—but don't *assume* that you will without working up to it. As you feel better, adjust your dose downward according to the Pulse Method, described in chapter 12.

- If you want more information about any of the procedures or supplements mentioned in this chapter, consult the Resources section at the end of this book. See also my website, www.DrBenLynch.com; it too has a Resources section.

- If you find that you have a single dirty gene, go right to that gene and follow the protocol for Spot Cleaning it. Even if you scored only a 1 for that gene, it might need a short, quick Spot Clean. As always, tune in to how you're feeling and adhere to the Pulse Method when it comes to supplements.

- If you find that you have multiple dirty genes—and most of us do— you might assume that you should address your dirtiest gene first. I've found that that's not the most effective approach, however. Instead, you should Spot Clean your genes in this order:
  —DAO
  —PEMT
  —GST
  —COMT
  —MAOA
  —MTHFR
  —NOS3

Implement the Pulse Method

As we discussed in the Soak and Scrub chapter, it's important to implement the Pulse Method to fine-tune your personal dosage of supplements. If you need a refresher, please go back to page 218.

Let me give you a couple more examples showing how the Pulse Method can help.

People who feel depressed often take supplemental methylfolate. In a few days, they feel great! Then they begin feeling irritable, snappish, wired; they experience that "jumping out of their skin" feeling. Oops—they need to cut their methylfolate dose right away. The Pulse Method could have helped them figure out how much to take without the seesaw effects described here. Because your body is always changing, the "right amount" of any supplement is always changing too.

Here's an example of you using the Pulse Method for the first time. See if you can spot where you went wrong.

Suppose you have a clean DAO gene and you want to start supporting your dirty PEMT with phosphatidylcholine. You evaluate how you're feeling: slightly anxious, a bit constipated, and a little achy in the muscles. You know that phosphatidylcholine may help support you in all these areas.

You start by taking one capsule with breakfast. For a few days, you don't feel anything, but on the fourth day, you notice that you're more calm and you're going to the bathroom a bit better than usual. You keep improving gradually over subsequent days, and your symptoms seem to be basically gone on the twentieth day. You're so stoked! You keep taking the phosphatidylcholine because it's working so well for you.

On the thirtieth day, you remember to tune in before you take your supplement. That day you sense a change: you realize that you seem a bit depressed. You reach for the bottle and think, "Wait. I was anxious, then I felt great for about two weeks, but now I feel depressed. I need to stop taking this supplement for now. I'll take it again if I feel I need it—if I get a bit anxious or constipated, or if my muscles hurt."

Did you catch it? Overall you did great here but you made two mistakes and it cost you. The first mistake occurred on the twentieth day—you felt great! What should you have done? You should have stopped taking phosphatidylcholine yet you continued taking it. Now you had an excessive amount of this supplement and it made you feel depressed. The second mistake occurred for about ten days—you were not tuning in each day. You remembered to do so on the thirtieth day. At the end, you figured it out and now know you should stop taking it when you feel great and to resume when you need it.

In the beginning, the Pulse Method will be a learning curve. As with anything new, daily practice is what will make it routine and easy. I'm confident you'll get it and experience the tremendous benefits!

As you go through Spot Cleaning, make sure to tune in to how you're feeling. Apply the principles of the Pulse Method with supplements, and you'll greatly improve your outcome.

Spot Cleaning Your DAO

The Dirty DAO Lifestyle

- Select Clean Genes recipes that support your DAO.

- You may need to find a health professional to help you identify infections and heal your leaky gut. We'll try a few things together first.

- Finding a health professional who specializes in visceral manipulation can be a game-changer. Have him or her focus on your gallbladder, liver, and diaphragm. For more information, see PEMT Spot Cleaning (the next section, below).

Fixing High Histamine in the Gut

High histamine in the gut can be triggered by a variety of causes—pathogenic bacteria, leaky gut, and many more—each of which has its own distinctive fix. Let's look at those causes in turn:

- **Overgrowth of pathogenic bacteria.**
  —*Blastocystis hominis, Helicobacter pylori, Clostridium difficile,* and other bacteria are very common. Interestingly, if one person in your family has this type of pathogen, typically everyone else does, too. Using natural antimicrobials (see below) can help get rid of the pathogens, but they might come back if you're stressed, have low stomach acid, use antacids, take antibiotics, or consume contaminated food or water. Effective antimicrobials include olive leaf extract, mastic gum, oregano oil, wormwood, neem, black walnut, garlic, and ox bile. It is best to rotate them rather than using a blend of them every day. This helps prevent resistance.

—If you have gut pathogens, and most of us do, you should experience gas and bloating when taking an effective antimicrobial. Starting with a low dose after dinner is recommended. That way you're less likely to have a large "die-off reaction"—all those bacteria dying at once can make you feel horrible. And if you do have the common die-off reaction—gas and bloating—you'll have it while you sleep, which is easier than having it when awake.

—Use gas and bloating as your guide to tolerance. If you take one capsule of an antimicrobial and don't experience any gas or bloating, increase the dose the next evening. If that still doesn't do it, stop using that product and switch to another one.

—*Saccharomyces boulardii* is a beneficial yeast that helps eliminate harmful pathogens. You can take it one hour after taking antimicrobials. Because *Saccharomyces boulardii* isn't killed by antibiotics, it's a great probiotic to take while taking antibiotics. You should take it for only about three to six months and then stop, however. Restart it only if you begin taking antibiotics or have a particular need, such as a gut reinfection.

—If you're not seeing results, see Appendix A for lab tests that can help you determine which pathogens you have and what will kill them.

—Restore your gut with probiotics after you've worked on eliminating pathogens. Consider replenishing with a blend without *Lactobacillus* first, such as a blend of *Bifidobacterium* probiotics. As with antimicrobials, after dinner is the best time to take probiotics.

—If you have significant gut problems, work with a health professional.

- **Leaky gut and gut inflammation.** Either leaky gut syndrome or an inflammatory condition such as ulcerative colitis or Crohn's disease contributes to a dirty DAO. None of these conditions will heal if you're stressed, eating foods to which you're intolerant or allergic, and/or have an overgrowth of pathogenic bacteria, yeast, or parasites.

—After you've worked on eliminating the pathogens, consider using L-glutamine powder to heal your small intestine, which is where your DAO enzyme lives. If your small intestine is unhealthy, your DAO's home may be in need of a remodel. Help your DAO enzyme by repairing its house. Start small, with 1 gram of L-glutamine powder. This supplement can increase irritability in some people. If that happens to you, stop using it for a couple days while you take some magnesium, vitamin B$_6$, and niacin. Keep taking those supplements while you resume the L-glutamine.

—A more effective option is to use a combination of L-glutamine, aloe vera, zinc carnosine, and marshmallow root.

- **SIBO.** Small intestinal bacterial overgrowth is associated with many causes, including antibiotic use, antacid consumption, constipation, low serotonin, sluggish bile flow, a diet high in refined foods, and excessive probiotic supplementation. Identifying the cause of SIBO is a must, or it will come right back after every attempt to treat it.

    —Ox bile in small doses can help support the elimination of harmful bacteria in your small intestine, which will also support your DAO. Start with 250 milligrams at dinner.

    —See PEMT Spot Cleaning (below) to get your bile moving again, which often helps eliminate SIBO.

- **A system that's too acidic.** Your DAO likes certain conditions. If your intestines are too acidic, DAO won't work well. If that's your problem, taking digestive enzymes and betaine HCL may help support your dirty DAO. The betaine HCL triggers your pancreas to secrete enzymes that reduce the acid in your small intestine. *Take only with meals.*

- **Food and drink high in histamine.** Lower your consumption of dietary histamine (see page 122) until you've healed your digestion and gut. Once you've healed your digestive tract by eliminating pathogens and providing the nutrients it deserves, you may find that you're able to eat histamine-containing foods again.

    —Beverage choice is especially important. The histamine in drinks or produced in response to them can overwhelm your DAO enzyme, creating such symptoms as headache, runny nose, itchy skin, tingling sensation, sweating, fast heart rate, and irritability. Reevaluate your consumption of the following beverages:

—*Juices and citrus.* Greatly reduce or completely eliminate from your diet drinks that contain citrus.

—*Champagne and wine (especially red, but even white can be an issue).* If you get headaches from wine, you may be experiencing sulfite sensitivity, discussed in chapter 9. Because sulfites interfere with absorption of vitamin B_1, which you need for many functions, it's no wonder they make some people feel bad. If you find yourself sensitive to sulfites, consider taking the supplement molybdenum. Look for molybdenum that isn't bound to ammonia, as many are. Common capsule dosages range from 75 to 500 micrograms. If you get molybdenum in liquid form (at 25 micrograms per drop), you can experiment to see what works best for you. Many people are sensitive to sulfites even if they don't know it. Trying some molybdenum early on may provide some incredible benefits. Just be aware that any supplement comes with potential side effects, and more is not necessarily better. If you take a lot of molybdenum for too long, it can drive up your uric acid levels and cause conditions such as gout. If you start to experience any negative effects, stop taking molybdenum and add in *pyrroloquinoline quinone,* commonly called PQQ. PQQ will help reduce the side effects from too much molybdenum.

—*Lime juice, tomato juice, and cocoa drinks.* These can also put you over the edge with their load of histamine. You may be able to tolerate an ounce or so, and as you improve you may find yourself tolerating more and more. For now, though, be careful. Symptoms can appear rapidly—anywhere from within seconds to half an hour.

—Histamine-containing foods aren't as critical as beverages. Some people are able to tolerate a small amount of such foods, but a full serving puts them over the edge. Symptoms can be delayed, especially with food, so keeping a food journal is key. The app CRON-O-Meter or other programs can help you identify which foods you can tolerate.

- **Undergrowth of bacteria that break down histamine.** If undergrowth of histamine-tackling bacteria is the cause of your dirty DAO, you need to take probiotics to replenish those bacteria while avoiding the probiotics that might make your condition worse.

—A combination of *Bifidobacterium* and *Lactobacillus plantarum* probiotics is fantastic at helping break down histamine.

—Avoid *Lactobacillus* probiotics until you restore your gut, including *Lactobacillus casei* and *Lactobacillus bulgaricus.*

- **Medications.**

—Metformin slows the DAO enzyme, thereby increasing histamine. However, stopping this medication is likely not an option for people it's been prescribed for. If you're one of those people, the key is to understand that you may be more histamine-intolerant because of the medication, and that you should therefore reduce your intake of histamine-containing foods and drinks.

—Aspirin and other NSAIDs and salicylates also contribute to increased histamine release. Instead of relying on these anti-inflammatory medications, look for natural ways to reduce inflammation. Low-dose naltrexone (LDN), a prescription medication, is quite well tolerated by many. Also, since inflammation is commonly associated with chronic infections, have your doctor look for them.

Additional Supplements for Your DAO

- **Copper.** The primary nutrient that your DAO enzyme needs to work properly is copper. Consider trying a supplement that contains that nutrient. Most people get all the copper they need quite easily from foods, but if you've been taking zinc supplements for a while, you might have developed a copper deficiency. For a list of copper-containing foods, see page 129. If you decide to go with a supplement, start with a low dose, because copper can be inflammatory. Consider, for example, 1 milligram of copper with a meal—but only if it's not already in your multivitamin. (See Appendix A for lab tests that can measure your copper levels.)

- **Histamine blockers.** A combination of stinging nettle, luteolin, bromelain, and quercetin works wonders to help keep histamine locked up and not troubling you.

- **Vitamin C and fish oil.** These nutrients help stabilize mast cells (cells that store and release histamine).

- **Cell membrane supporters.** Healthy cell membranes are needed to keep histamine inside individual cells. For ways to support your cell membranes, read Spot Cleaning Your PEMT, below.

- **Buffering agents.** Sodium bicarbonate and potassium bicarbonate can be lifesavers if you eat acidic foods or are having a histamine reaction. Simply take a capsule or two with filtered water. The beneficial results are often immediate.

Spot Cleaning Your PEMT

The Dirty PEMT Lifestyle

- Select Clean Genes recipes that support your PEMT.

- Understand that you'll need additional support during pregnancy and breastfeeding.

- You may also need additional support after menopause.

- Consider visceral manipulation of your liver, gallbladder, and diaphragm.

Supporting Low Estrogen

- If your premenopausal estrogen levels are low, you need to get help balancing them from your health-care professional.

- Common reasons why estrogen can be low:
  —High stress that uses up precursor hormones for cortisol instead of estrogen production.
  —Weak fat absorption that translates to low cholesterol, which in turn results in low estrogen.

Supplements for Your PEMT

- **Phosphatidylcholine.** Support your cell membranes with phosphatidylcholine. Use a non-GMO, soy-free form, because soy is a common allergen; furthermore, most soy is GMO. Store liquid phosphatidylcholine in a cool, dry area, but not in the refrigerator (which would make it harder to pour). If you're not a vegan or vegetarian, you can also find phosphatidylcholine in gelatin capsules. Taking phosphatidylcholine supplements can lead to feelings of depression, so be sure to follow the Pulse Method (again, see chapter 12) and fine-tune your dosage.

- **Creatine.** Take creatine in order to conserve SAMe so that there's more SAMe available to help make needed phosphatidylcholine.

Spot Cleaning Your GST/GPX

The Dirty GST/GPX Lifestyle

- Select Clean Genes recipes that support your GST/GPX.

- Avoidance is king. Clean up your environment, and limit your exposure to chemicals that you touch, breathe in, or ingest.

- Sweating via a sauna, an Epsom salt bath, exercising, or hot yoga helps your body expel the industrial chemicals that burden your GST/GPX.

- Eating fiber supports detoxification as well as binding and removing xenobiotics. It also bolsters the beneficial bacteria that support detoxification.

- Dry-brushing your skin and massage are fantastic ways to help support detoxification.

Supplements for Your GST/GPX

- **Liposomal glutathione.** This easily absorbed form of the supplement helps deliver glutathione directly into your cells so that they can bind to the compounds. Start slowly and work up. I recommend that you skip some days; consider taking glutathione a few times a week rather than daily. If you notice it helping, then move to daily and adjust as needed.

- **Riboflavin / Vitamin B$_2$.** You need this nutrient to regenerate damaged glutathione back into useful glutathione. Otherwise, your glutathione remains damaged and can contribute to further cell damage.

- **Selenium.** Without selenium, you can't use your glutathione to get rid of hydrogen peroxide. You can have all the glutathione you want—but without selenium, it's "stuck."

- **Detox support powders.** There are a variety of detoxification support products available. If you use a powdered detox supplement, you can add it to a smoothie for a fast and easy breakfast or lunch.

Spot Cleaning Your Slow COMT

The Slow COMT Lifestyle

- Select Clean Genes recipes that support your slow COMT.

- Be fully aware that when you get stressed, it can take you some time to calm down. Allow sufficient time to recover from the aggravating factors. Find what works for you: walking away, breathing exercises, and stepping outside are some useful tactics.

- Do your stimulating activities earlier in the day and the calming ones in the evening. Exercising, gaming, and dancing can all be stimulating to the point where they interfere with your sleep. This sets you up to fail the next day. Simply adjust when you do your activities and succeed!

- You're a thinker. Find activities that stimulate your brain or you'll be bored.

- While you're a thinker, you need to practice calming activities such as hiking, meditating, playing or listening to music.

- Work hard. Play hard. Understand that you do tend to be a workaholic and that is okay as long as you balance it with extended days off and vacations. If you don't balance it, you'll end up stressed, aggravated, and burned out. It's incredibly important you balance your tendency to overwork. Plan your vacations just as you plan your workday. Get vacations on the calendar.

- Identify your routine stressors and remove as many as possible. News, particular "friends," a long work commute, routine chores you can delegate to your kids or professional services (housecleaning, dishwasher loading/unloading, cooking).

- Sleep is a tough one for you. Night owl all the way. You do your best work in the evening, as it is quiet, people don't bug you, and you're incredibly productive. The issue is you're trashed the next day, which sets you up for being more emotionally reactive. Find ways you can do your best work earlier in the morning, before people wake up. I know this sounds awful now, but as you switch over you'll be amazed at the difference in your productivity, health, and mood.

- Consider relaxing yet healthy activities such as massage, Epsom salt baths, and saunas. While these are fantastic for everyone, you really need them in order to stay on top of your game without burning out—or burning up!

Supporting a Slow COMT

- Optimize your weight, because body fat is estrogenic. If you can't lose weight, you may have a dirty GST/GPX.

- Use cosmetics that are low in phthalates and other compounds. Buy organic produce. Use the lists produced by the Environmental Working Group to figure out which items are most important to buy organic (www.ewg.org/foodnews/dirty_dozen_list.php).

- Eat more beets, carrots, onions, artichokes, cruciferous vegetables (broccoli, cauliflower, kale, brussels sprouts, cabbage). If you get gas from these vegetables, consider taking the mineral molybdenum.

- Support your liver with bitter vegetables such as dandelions and radishes.

- Limit high-catechol foods and drinks, and monitor caffeine.
  —As we've seen, catechols are found in green and black tea, coffee, chocolate, and a few green spices such as peppermint, parsley, and thyme. You don't have to *eliminate* them; just be aware of how they're affecting you and limit as necessary, especially during bouts of PMS or insomnia, when you may want to avoid them completely. If you struggle with insomnia, have your green tea in the

morning. If you're getting close to menses and are starting to feel irritable, don't drink a ton of green tea at this time—maybe just have a cup and see how you feel. You get the picture. Moderation is key. Absolute avoidance is hardly ever needed. Just be aware; listen to your body.

—Watch your intake of caffeine, which can make you edgy and deplete your magnesium.

- Limit excess histamine. If your histamine is high, then you'll be relying on methylation to process it. Learn how to reduce your histamine levels by reading Spot Cleaning your DAO, earlier in this chapter.

- Limit your protein intake.
  —Protein provides tyrosine, a nutrient that your COMT enzyme uses. If you give it a lot of tyrosine, you're potentially slowing it down. If you're following a high-protein GAPS or Paleo-style diet and feeling anxious, it could be because you're consuming too much tyrosine, which is fueling your already likely higher than normal dopamine.
  —Eat your biggest portion of protein at breakfast, have a moderate amount at lunch, and have very little at dinner. This way you'll focus well and be "on" during the day, and in the evening you'll be able to wind down.

- Be aware of medications and supplements.
  —ADHD, SSRI, and thyroid medications may make you feel even more on edge, so be careful with them. Talk with your doctor if you're experiencing side effects of insomnia, irritability, elevated estrogen levels, or histamine issues.
  —Steroids can increase stress and thus increase demand on your COMT, slowing it further.
  —Tyrosine can be stimulating, which increases your anxiety and thus raises pressure on your COMT enzyme. Absolutely do not take tyrosine-containing supplements within six hours of bedtime.
  —Methylfolate supplements can increase nitric oxide, which in turn stimulates dopamine release and potentially slows your COMT. You often need to open up your slow COMT before supporting it with methylfolate.

—L-dopa can make too much dopamine and push on COMT, again slowing it down.

—Bioidentical estrogen hormones can slow your COMT.

—Birth control that contains estrogen can also slow your COMT.

- Evaluate your thyroid function.

  —Oral estrogen hormone replacement can lead to hypothyroidism. The estrogen stimulates the production of a protein called thyroid-binding globulin (TBG), which carries your thyroid hormone. As a result, too much thyroid hormone becomes bound— but only the unbound, free version of the hormone is active. Even if your total blood levels of thyroid are normal, the amount of active thyroid in your system may be too low. To evaluate thyroid function, it's not enough to check your TSH, which is what most doctors do. You also have to check free T4, free T3, reverse T3, thyroid antibodies, and TBG.

  —Estrogen is far from the only influence on thyroid function, so be sure to check Appendix A to find out which tests to order for evaluating your thyroid.

Supplements for Your Slow COMT

- **Adaptogens.** Use adaptogens as described in Soak and Scrub (chapter 12).

- **Magnesium.** A surprising number of people are deficient in magnesium. You should be getting some magnesium from your electrolytes, as mentioned in the Soak and Scrub chapter. If you want to supplement additionally with this mineral for its calming effect, magnesium glycinate chelate is a good form; it helps moderate anxiety and supports liver fiction. Three other effective forms are magnesium taurate, magnesium malate, and magnesium threonate.

- **Taurine.** If you take high-quality magnesium supplements and still can't get your magnesium levels high enough, you may have low levels of taurine, a mineral that helps with magnesium absorption. Low taurine levels are caused by many things, but one common reason is *gut dysbiosis*—an imbalance of bacteria in the gut. Spot Clean your

DAO to help correct this issue. Consider working with your doctor to evaluate your digestive function with a comprehensive digestive stool analysis (CDSA). If you're able to right your bacterial balance, you'll support your taurine levels at the same time—and thus your magnesium levels will normalize as well.

- **SAMe.** This can be a very helpful supplement, but only if your Methylation Cycle is working well. To find out, take a 250-milligram capsule of SAMe before bed. If it helps you fall asleep, great. Keep using it. If it makes your insomnia worse, then you might be low in methylcobalamin and/or methylfolate, or your Methylation Cycle might be blocked by heavy metals, insufficient glutathione, excess hydrogen peroxide, or some other factor. If insomnia worsens, stop taking the SAMe until your Methylation Cycle is back in balance—but meanwhile, if you're now wide awake and staring at the ceiling, you can neutralize that insomniac effect by taking 50 to 150 milligrams of niacin. This will help break down the SAMe you just took and get it out of your system.

- **Phosphatidyl serine.** This supplement can be a very helpful sleep aid, especially in conjunction with magnesium malate, niacin, and vitamin B_6.

- **Creatine.** When your body makes creatine, it uses up the majority of your methyl donors—those nutrients that support methylation. When you take supplemental creatine, you conserve methyl donors and SAMe, leaving your SAMe free for other things, like helping your slow COMT. Creatine has helped a number of people who are unable to take methylfolate, methylcobalamin, or other methyl donors. It is safe and well tolerated by many who are otherwise sensitive to supplements. Autistic children or those who are slow to speak do exceptionally well with creatine. We're seeing children who have never said a word begin speaking with creatine supplementation. Be sure to drink a glass of filtered water when using creatine. I also often recommend mixing creatine and electrolytes with filtered water and drinking it from a water bottle or Thermos throughout the day and before exercising.

- **Phosphatidylcholine.** Supplemental phosphatidylcholine is useful for conserving SAMe since, like creatine, the phosphatidylcholine your body produces uses up a lot of SAMe. Taking additional phosphatidylcholine leaves your body more SAMe to support your COMT. Be sure to use a non-GMO, sunflower-based supplement.

- **Indole-3-carbinol and DIM.** These supplements help break down estrogens so that they can be eliminated from your body. Often they come packaged together.

Spot Cleaning Your Fast COMT

The Fast COMT Lifestyle

- Select Clean Genes recipes that support your fast COMT.

- Engage in activities that stimulate and engage your brain. Good choices are playing music, dancing, singing, participating in debate club, hiking in groups, playing team sports, and doing other social activities. Playing a solo sport that keeps you focused—tennis or martial arts, for example—is also beneficial.

- Go for a run or exercise in the morning. This is excellent for you: it gets your blood flowing and boosts your dopamine right away. Find some way to be physically active *every* morning, even if it's only parking your car further away from work or walking to the barista stand for a cup of tea before you start work. Consider more berries, green tea, and flavonoids to slow down the burn of your estrogen and dopamine.

- Notice how your temper works. You may find that you sometimes get involved in arguments, or you may notice that one of your fast-COMT kids tends to instigate fights. Fights spike dopamine, and if you have a fast COMT, the rise in dopamine makes you feel better. So let's raise dopamine by eating protein rather than picking fights! I credit Dr. Daniel Amen for pointing this out to me years ago.

- Be self-aware. Understand that you may naturally bounce from one thing to another. The key is devoting enough time to each activity that it's meaningful and you accomplish something. Work hard on one thing for thirty minutes or so, shift your focus to something else for another half hour, then come back to what you were doing before. This way you have the variety you crave, yet you also accomplish things.

- Addicted? Be aware that you're susceptible to spending way too much time on social media, video games, shopping, TV, and many other activities. Take it as a warning sign that you need to support your fast COMT with the tools in this book.

Supporting Your Fast COMT

- Ensure that you're absorbing the protein you consume. Follow the Soak and Scrub precepts faithfully. See Spot Cleaning Your DAO (earlier in the chapter) to heal your gut. If you're still struggling with absorbing protein after doing those things, an amino acid blend can be very helpful. Capsules are best, because amino acid blends taste awful.

- Make sure you're getting enough protein at every meal. You need a good supply of protein to keep you focused.

- Be aware of medications and supplements.
  —SAMe. Taking some SAMe via the Pulse Method can be useful if your fast COMT suddenly becomes slow in response to your new supplements and lifestyle. Be careful with this supplement, however; taking it daily may lower your dopamine and norepinephrine, leaving you feeling flat or depressed.
  —Phosphatidylcholine and creatine. These supplements might be okay for you, but if you notice that you're feeling a bit more depressed than usual, you might need to evaluate your protein intake and increase your dopamine levels. See also Spot Cleaning Your PEMT (earlier in this chapter) for the discussion of potential side effects from phosphatidylcholine.
  —Estrogen-containing birth control or estrogen bio-identical hormone. If this type of birth control or bio-identical hormone improved your mood and focus, the estrogen may have slowed your fast COMT. Talk with your doctor about checking your estrogen levels. (See Appendix A.)

Supplements for Your Fast COMT

- **NADH.** If you're slow to wake up in the morning, consider NADH with CoQ10. These two compounds supply your mitochondria immediately with fuel allowing them to produce your cellular energy, ATP. Typically, your body makes NADH through a long, involved process. You completely bypass this process by taking these. Take one tablet and let it dissolve under your tongue while you're still lying in bed. This can literally wake you up in minutes. If you're trying to quit caffeine—coffee or energy drinks—it's a great nonstimulating replacement. The NADH with CoQ10 provides clean sustained energy compared to the spike and crash from caffeine. Never take with food. Always take upon waking in the morning or at least one hour away from food.

- **Adrenal cortex.** If you can't wake up in the morning or you feel that you're just dragging through the day, adrenal cortex can be a huge help. Adrenal cortex supports your body's ability to make the hormone cortisol. Those who have chronic stress may have lower levels of cortisol. Adrenal cortex helps us wake up as it is cortisol that helps us wake up in the morning. Take one 50-milligram capsule with breakfast. It's a potent supplement, so definitely fine-tune your dosage via the Pulse Method. You may find that you need to take it only a few times a week.

- **Tyrosine.** This supplement—a precursor to the neurotransmitters dopamine, norepinephrine, and epinephrine—can be great for you, especially when taken in the morning and early afternoon. Do *not* take it within six hours of bedtime, though.

- **5-HTP.** While this supplement—a precursor to the neurotransmitter sertotonin—is mainly used for those with a fast MAOA, it may also be useful for those with a fast COMT. If you have a slow MAOA, be cautious. Higher serotonin levels slow a fast COMT, which is why I recommend that people with a fast COMT and a fast MAOA consider 5-HTP. However, do *not* take 5-HTP if you're on an SSRI.

Spot Cleaning Your Slow MAOA

The Slow MAOA Lifestyle

- Select Clean Genes recipes that support your slow MAOA.

- The recommendations for a slow COMT (earlier in this chapter) might also benefit you greatly, since both slow genes reduce how fast dopamine and norepinephrine clear out of your system.

Supplements and Medications That May Adversely Affect Your Slow MAOA

- **SSRIs.** If you're experiencing headaches, irritability, insomnia, discuss with your doctor that you believe the dose is too high or the medication is potentially not suitable for your genes.

- **Testosterone.** Supplemental doses of this hormone can increase aggression, especially in people with a slow MAOA. Ask your doctor to reevaluate your testosterone dose and keep it as low as is medically necessary.

- **Thyroid medication.** This type of medication can also increase aggression and anxiety in someone with a slow MAOA. If you experience such symptoms, talk with your doctor about adjusting your dose.

- **Tryptophan, 5-HTP, and melatonin.** Consider stopping these supplements. If prescribed, discuss them with your doctor. All of these put pressure on your MAOA and slow it down.

- **Tyrosine.** This supplement can put a burden on both your COMT and your MAOA genes and slow them down, so reduce or cut out your dose; discuss with your doctor if they've been prescribed.

- **Inositol.** Like lithium orotate, inositol helps regulate serotonin. However, you can inadvertently burden and slow your MAOA with this supplement. Lithium and inositol act opposite to each other, so if you respond poorly to one, you should respond favorably to the other.

Supplements for Your Slow MAOA

- **Riboflavin.** Consider taking 400 milligrams of riboflavin to help support your slow MAOA.

- **Lithium.** Consider 5 milligrams of lithium orotate, a supplement that helps calm the activity of excess serotonin.

Spot Cleaning Your Fast MAOA

The FAST MAOA Lifestyle

- Select Clean Genes recipes that support your fast MAOA.

- Identify potential causes of inflammation and work to eliminate them. Typical causes of inflammation are diet (see next bullet point), poor sleep, stress, chemical exposure, and improper breathing—all discussed in the Soak and Scrub chapter.

- Identify inflammatory food allergies and food intolerances. Lab testing is great at identifying food allergies but not very accurate at identifying food intolerances. Consider following an elimination diet to find out more.

- Make sure you're not overtraining. Evaluate your exercise by measuring your heart rate variability (HRV) using apps such as HRV4Training or the ŌURA ring. Don't train hard if your HRV drops a lot or if your ŌURA ring suggests that you should take it easy.

- Mold is a common trigger for MAOA issues. Call an environmental inspector to come and visit your home or office and assess. Your car, camper, or boat may also be harboring mold.

- Infections are another common trigger. They're tough to spot, though—even for health professionals. If you're struggling with a fast MAOA, consult a naturopathic physician or integrative/functional medicine doctor who specializes in chronic infections to determine whether you have an undiagnosed infection. In the meantime, continue the Soak and Scrub and try the supplements suggested below, which can support you while you're fighting off an infection. See Spot

Cleaning Your DAO (earlier in this chapter) for ways to help you eliminate pathogens.

Supplements for Your Fast MAOA

- **NADH.** If you're slow to wake up in the morning, take NADH with CoQ10, as was suggested for a fast COMT. Dissolved under your tongue while you're still in bed, a tablet wakes you up in minutes. As was noted earlier, this is a great wake-up solution if you want to quit caffeine.

- **5-HTP.** At 50 milligrams per day, this is typically an effective supplement for a fast MAOA. If after a couple of weeks you don't notice enough improvement, try a larger dose. If you're not staying asleep at night, consider a sustained-release capsule to provide small amounts of 5-HTP continuously throughout the night. However, do *not* take this supplement if you're taking an SSRI.

- **Inositol.** Start with a small dose to regulate serotonin and improve mood, and increase as tolerated.

- **Melatonin.** This supplement might help you sleep at night.

- **Liposomal curcumin.** Consider taking this great anti-inflammatory one to three times daily. This helps slow the Tryptophan Steal that we discussed earlier, thereby conserving tryptophan for your fast MAOA.

Spot Cleaning Your MTHFR

The Dirty MTHFR Lifestyle

- The Soak and Scrub should cover all your bases with this gene.
- Select Clean Genes recipes that support your MTHFR.

Hypothyroidism and a Dirty MTHFR

- Hypothyroidism slows your ability to activate vitamin B_2, so talk with your doctor about evaluating your thyroid function. (Also see Appendix A.)

- Support your thyroid function by reducing stress, supporting your adrenals, healing your gut, avoiding chemicals, filtering your water, getting adequate sleep, and fighting off infections.

- See Spot Cleaning Your DAO and Spot Cleaning Your Slow COMT (earlier in this chapter) for additional support.

Supplements for Your MTHFR

- **Riboflavin / vitamin B$_2$.** This is a nutrient that MTHFR needs to work properly. The most active form is riboflavin-5-phosphate (R5P). A daily dose of 20 milligrams is typically enough for most people; however, as much as 400 milligrams may be needed for others, especially those struggling with migraines.

- **L-5-MTHF or 6S-MTHF.** These are both quality forms of methylfolate. Many people do well with just a multivitamin containing 400 micrograms of MTHF. If you feel no change with 400 micrograms, try more. However, don't make a huge jump; try doubling the dose. Many health professionals go straight to higher doses of 7.5 milligrams or above. While this might provide initial benefits, it can cause significant side effects within days. Because this nutrient is so powerful, tuning in to your body as you implement the Pulse Method is of utmost importance. Another option is to use liposomal MTHF. That way you can regulate the dose and deliver the MTHF right inside your cells. (See the Resources section on my website.)

- **If you're taking 5 milligrams or more of methylfolate and don't notice a response, one of these might be the reason:**
  —You have folate receptor antibodies and they're blocking your folate receptors. (See Appendix A for more about tests that can determine whether this is true.)
  —You're still consuming folic acid, and it's blocking your receptors.
  —You're deficient in vitamin B$_{12}$ so your methylfolate is trapped and can't be used.
  —You're using an inferior supplement containing D-methylfolate instead of L-methylfolate. If the supplement does not specify L-methylfolate or 6S-methylfolate, it may have the inferior

D-methylfolate form. Your body does not use D-methylfolate. Ask the manufacturer.

—Your Methylation Cycle is blocked for other reasons, such as heavy metals, oxidative stress, infections, or medications.

Caution: If you experience anxiety, irritability, runny nose, joint pain, insomnia, or hives, you may be taking too much MTHF. Stop taking it immediately and take 50 milligrams of niacin every twenty minutes, until your side effects disappear (for a maximum of three times). However, if you have low blood pressure of 90/60 or lower, be careful: the niacin might further lower your blood pressure.

Spot Cleaning Your NOS3

The Dirty NOS3 Lifestyle

- The Soak and Scrub should take care of most of your dirty NOS3.

- Select Clean Genes recipes that support your NOS3.

- Keep your GST, PEMT, MTHFR, COMT, MAO, and DAO clean, and your NOS3 will pretty much take care of itself. This is why NOS3 is the last gene for you to clean up if others are dirty. Typically, it's those other dirty genes that are causing your NOS3 to become dirty. Tackle them one by one, and you'll see results. Don't rush it.

- Make sure you're doing some form of exercise, even if it's just a brisk daily walk. Exercise stimulates your NOS3 to function. However, don't overdo it, as too much exercise can uncouple your NOS3. (Uncoupling is discussed in chapter 10.) You can tell you've done too much when you have prolonged soreness for a day or two after exercising.

- Good breathing is important for you. Seriously consider daily yoga or tai chi and breathing exercises. Pranayama, the science of breath, is a good option.

- Sauna is great at stimulating your NOS3—especially on a twice-weekly basis—so give that a try. Don't push it!

Supplements for Your NOS3

If you're inflamed, have high homocysteine, or are fighting a known infection of any type, I recommend first reducing your homocysteine and fighting the infection before supplementing for NOS3. In addition, make sure to clean your other dirty genes before addressing NOS3.

- **Ornithine, beet root powder, or citrulline.** If you're generally healthy, increasing arginine levels via these supplements may be all you need. (I'm not a fan of directly supplementing with arginine, as you read in chapter 10.)

- **PQQ.** This is a must to keep your nitric oxide healthy and keep it from turning into superoxide. If you're exercising hard or tend to get significant postworkout soreness, take one of these capsules after working out. Those with fibromyalgia or chronic fatigue should do very well using PQQ.

- **Liposomal vitamin C and liposomal glutathione.** These supplements help keep your nitric oxide happy and prevent it from turning into superoxide.

Now What?

Perhaps you say, "I'm better, but I'm still not where I want to be. Now what?" Good question.

You've followed the Soak and Scrub. You're *living* it.

You've worked hard on Spot Cleaning your genes.

Yet you continue to struggle.

If that's the case, I recommend finding a functional/integrative practitioner: a licensed naturopathic physician, functional medicine professional, or environmental medicine professional. These health professionals work hard on identifying the cause of disease rather than suppressing the symptoms. (See Appendix A for information on available lab tests.)

By adopting the Clean Genes Protocol, you've come a long way in getting the basics done. Your health professional will have worked with you on these basics. Now you can work with him or her to dig deeper, seeking out both hidden infection and hidden chemical exposure.

- **Identify hidden infections.**
  —**Mouth.** Root canals, diseased gums, and throat are all frequent sites of infection. If you have bleeding gums, bad breath, or bad teeth, you likely have a persistent infection in your oral cavity or a chronic infection elsewhere that's causing poor dental health. Work with a biological dentist—that is a dentist who takes a whole-body approach—to fix this problem.
  —**Nose.** Your nose is a frequent site of mold and infection. Ask your doctor to swab your sinuses and nostrils to check for infection, especially if you have chronic sinus issues of any type.
  —**Gut.** Even if you don't have digestive issues, you might have full-body symptoms from imbalanced bacteria. Ask your health professional to order a comprehensive digestive and stool analysis (CDSA) to find out.
  —**Blood.** Get bloodwork done to find out how your immune system responds to various pathogens. That will help you identify any viruses or bacteria you may be harboring.
  —**Urine.** A urinalysis can provide insight into a recurring bladder infection, as well as immune system markers.

- **Identify hidden sources of chemical exposure.**
  —**Mouth.** If you have a lot of old fillings, you may need to discuss having them replaced with a less toxic substance by a biological dentist.
  —**Urine.** Your kidney is a wonderful filter. There are great lab tests out there that can quickly evaluate hundreds of chemicals via your urine; they can also identify heavy metals. Once you know your contaminants, you can tackle their removal.
  —**Blood.** Blood tests can identify heavy metals, carbon monoxide, and other problematic compounds that your doctor can then help you eliminate from your body.

You've done a tremendous job getting this far. Take it another step by working with a quality health professional to uncover these hidden issues and eliminate them one by one. You're well on the way to reaching your genetic potential!

CONCLUSION

The Future
of Gene Health

This book covers a far more advanced approach to genetics than is currently practiced in medicine today. The information you now have at your fingertips is not a Get Healthy Quick scheme. It's a lifetime tool that you can revisit any time you need to tweak your dirty genes.

Your daily life should consist of the Soak and Scrub. Mine does. This is not something you do for two weeks and stop. You did it for two weeks to prepare you for Spot Cleaning—and you did it *while* you were Spot Cleaning. Now that you've done both, you'll want to continue implementing the Soak and Scrub principles each and every day.

Yes, some days you'll eat more than you should, or stay up late watching movies and partying with friends. Awesome! Do it! Live! Just know that you've dirtied your genes and you'll need to nurture them back to health. The good news is that now you know how to do it.

Let's face it. Our genes will always get dirty—every day. Some days they will get dirtier than others, but *each* day they at least gather dust. Now you know how to brush that dust off daily with the Soak and Scrub so that you can avoid a huge spring cleaning.

Even while implementing the Soak and Scrub principles, significant life stressors, injuries, toxic exposures, and lifestyle changes can really dirty your genes. When that happens, revisit Laundry List 2 and Spot Clean as needed.

What's the Latest on SNPs?

Researchers continue their investigation fervently. Over the years they will surely find more and more SNPs, which will create more and more rabbit holes for us to go down.

Many people will say, "Oh, wow! That SNP caused all my problems! What supplement do I take for it?"

Now that you've read this book, you'll know how to answer. You'll say, "Look. SNPs have been around as long as humans have. The most important things to consider are lifestyle, diet, mindset, and the environment. You're right: a SNP absolutely can impact the function of a gene by slowing it down or speeding it up. However, even a single industrial chemical, such as mercury or aluminum, can impact hundreds of genes—and way more significantly than a single SNP or even a dozen SNPs."

Understanding the combined impact of SNPs and lifestyle factors on our genetic function is where we need to be. Unfortunately, too many of us are far from grasping this crucial linkage. These are the people who are randomly taking methylfolate to fix their MTHFR SNP or taking phosphatidylcholine to fix their PEMT SNP.

You're not taking that scattershot approach.

Instead, you're making the changes you need to lessen the work your genes have to do. You are:

- Breathing properly
- Sleeping deeply
- Exercising moderately
- Eating to nourish versus to satisfy your cravings
- Sweating
- Filtering your air

- Filtering your water

- Enjoying cleanliness as it should be—without a chemical scent

- Interacting with loved ones and friends

- Experiencing life

We're Just Beginning

Now that you've learned how and why your genes get dirty—and have the resources at your fingertips to clean them up—your next step is to take action.

You did it for two weeks to prepare you for Spot Cleaning—and you did it *while* you were Spot Cleaning. Now that you've done both, you'll want to continue implementing the Soak and Scrub principles each and every day.

If you haven't already adopted the Clean Genes Protocol, plan out when you're going to start the Soak and Scrub and get to it.

Today? Tomorrow? Next Friday?

I look forward to hearing your results and experiences!

I'm constantly researching, writing, presenting, and creating new resources to help you reach your genetic potential. You can keep in touch with my latest findings and with resources available at www.DrBen Lynch.com.

I love what I do. But it means absolutely nothing unless it gets implemented with great outcomes.

And so, I thank you. I thank you for investing in your health and taking the time to learn how to optimize your life. Without you, my work would mean nothing.

Someday, our paths may cross—or maybe they already have. Online on social media, in person at a conference, on an airplane or while hiking. If you see me, please stop me and share how my work has helped you or your family. *You* are the reason why I'm grinding away at the research. It is stories like yours that keep me going.

Share your stories with others as well. Many people are struggling and could greatly benefit from the knowledge you've learned within these pages. When you're talking with people and the conversation gets you thinking, "Man, his COMT is definitely slow," or "Sounds like she has a dirty MTHFR," reach out and give those individuals some tips to clean it up. They may listen, or they may not. What matters is that you've reached out and tried to help. Many times I have offered information, only to have it shot down. I've learned, though, that the key is simply planting the seed. In a few weeks or a couple of years, some person you gave tips to may stop you and say, "You know that thing you told me about genes? I looked into it a little further—and it's changed my life."

By helping others reach their genetic potential, we can all make the world a better place. Before you reach out and help others, though, I want you to reach up and put on your own oxygen mask first. Today is the day you're starting your journey to clean up your dirty genes. You deserve to reach your genetic potential. Now go for it!

Acknowledgments

Writing a book is hard. In fact, it was impossible for me—or so I thought. I tried (and failed) many times before producing these chapters. What changed was my support team. The turnaround started with my friend and colleague Peter D'Adamo. Without your introductions, Peter, this book would not exist.

Janis Vallely and Rachel Kranz: You two ladies are exceptional. Skill, patience, drive, persistence, dedication, and cheerleading were the much-needed attributes I got from you. With those attributes, you both helped me consolidate my scattered thoughts into an impressive book that will help many.

Julia Pastore: Thank you, and HarperCollins, for your trust in me. From our first conversation on the phone to this day, you've impressed me. Your direction, professionalism, and skills helped make *Dirty Genes* an exceptional book. I am truly forever grateful.

Adam Rustad: Thank you for taking the reins of Seeking Health, leading our team and freeing me up. Your skills and leadership provided me an incredibly important asset: time. Without it, this book would not have been written.

To my countless colleagues around the world who supported me on this journey: Thank you. We all know how new the field of genetics and epigenetics is. You are pioneers. I've taught you. You've taught me. Your dedication to helping your patients reach their genetic potential is staggering. That task is not easy. There's no guidebook. Your insights, patient experiences, and hunger for knowledge led me to create a starting point. I do believe that *Dirty Genes* will help make your tough—but rewarding—job a little bit easier.

To my former patients and clients: working with you has taught me so much. There were times where my recommendations caused grief for you. It was actually those moments that pushed me the most to figure out *why*. Together, we learned, moved forward, and got better.

To Mom and Dad: You were tough on me as a kid—and now I get why. Pushing me to figure things out on my own, work independently, and put in long hours on the ranch provided me the discipline and work ethic I needed—not only to succeed in life, however one defines that, but also to have what it takes to sit down and write this book. Thank you.

My boys: Tasman, Mathew, and Theo. Where to start with you three? I love you so much. I love that you push me. I love that you love me. I love that you understand that Dad has to take time away from you so he can help many others become and stay healthy—like you. I've learned a lot from you over the years. With the knowledge you've provided me, I'm able to help people all around the world. Thank you.

Nadia: We've shared many experiences—some tough and many exceptional. Writing this book pushed into some of our family vacations, but we still managed to make them fun. I love that you give me the freedom to dive in and work hard. I love that you're as passionate about helping others as I am. I love that you put up with my crazy ideas—"Let's try eating this way now," or "I think we should try X and Y, and stop Z." I know that these decisions affect not just me, but the entire family. It's through these many experiments that I learned much of what I know—and now we will help many others because of them. One thing is consistent: I love you and what we have together.

Lab Testing

The Clean Genes Protocol is a comprehensive program. The Soak and Scrub followed by Spot Cleaning should help your well-being significantly—if not completely! The beauty of *Dirty Genes* is that most of you can use this book without ever asking your doctor to order a special lab test. You can evaluate your own status using the two Laundry Lists, and follow the Clean Genes Protocol to optimize your health.

However, if you're struggling—if you've been "clean" for three or four months and don't see any improvement, or if improvement is proceeding at what seems to you a snail's pace—then you might need additional help. Ideally, you'll find a good integrative or naturopathic physician to help you, or you'll find a way to work with your conventional physician that addresses these issues. In either case, you'll then need labwork to evaluate your status.

Unfortunately, lab testing is something that can be either useful or, in the hands of a health provider lacking specialized knowledge, a big waste of money and time. If you get a baseline lab reading to see where you are now, lab testing can be useful going forward, as a gauge to see how you're improving with real data. However, most lab test results are expressed in terms of ranges that are based on the "average healthy person"—who is actually not very healthy! So you have to be prepared to see many of your labs come back reading "normal," and then a conventional doctor is going to insist that you're fine. Be aware that you can have plenty of problematic symptoms (or just not be at the peak of your health) and still get a "normal" reading. Without a knowledgeable provider, you can't rely on lab work to identify problems for you to repair.

Specialty testing, while potentially expensive, can be informative— but only if your health professional knows how to interpret the results. Basic labs are usually your best place to start, because they can provide you with some baseline information and are fairly affordable.

Ideally, though, you want to order all your lab tests at once. That way, you get a comprehensive picture rather than a scattershot presentation with results taken from different dates. Even within a week or two, your results might be affected by what you ate the day before, how stressed you're feeling, and other factors that, though subtle, can nonetheless have a big impact on your labs. Again, a skilled health professional is key, because he or she needs to put all the findings together and determine what the patterns mean. Unfortunately, few conventional MDs under- stand this approach to medicine, but a growing number do (including those I've trained!), as well as many naturopaths and functional/integra- tive medical providers.

Listed below are the lab tests I recommend. Most are far from stan- dard for a conventional practitioner, but an integrative or naturopathic physician will routinely order them. I've put the names of labs that offer these tests in parentheses after each test. The names and websites of these labs are provided at the end of this appendix.

I begin with a list of general lab tests. Then, in the remainder of the appendix, I move on to lab tests (including some from this general list) that are of particular benefit to each type of dirty gene addressed in this book.

General Lab Tests

- Complete blood count with differential (CBC with diff) (Quest Diagnostics, LabCorp)

- Thyroid panel: TSH, free T3, free T4, reverse T3, thyroid antibod- ies, TBG (Quest Diagnostics, LabCorp)

- Serum ferritin (Quest Diagnostics, LabCorp)

- Vitamin D: 25 OH vitamin D_3 and 1,25 OH vitamin D_3 (Quest Diagnostics, LabCorp)

- Lipid peroxidation (Quest Diagnostics, LabCorp)

- Fasting serum insulin (Quest Diagnostics, LabCorp)

- Glycated hemoglobin (HbA1c) (Quest Diagnostics, LabCorp)

- High-sensitivity C-reactive protein (hs-CRP) (Quest Diagnostics, LabCorp)

- Methylmalonic acid (Quest Diagnostics, LabCorp)

- Holotranscobalamin (Dr. Lal PathLabs)

- Advanced cholesterol panel (VAP) (Quest Diagnostics, LabCorp)

- Urinary organic acids (Quest Diagnostics, LabCorp, Genova Diagnostics, Great Plains Laboratory)

- Red blood cell (RBC) fatty acids (Doctor's Data, Quest Diagnostics, LabCorp, Genova Diagnostics)

- Chronic infections panel: viral, bacterial, Lyme, parasite, mold (DNA Connexions, Full View test; LabCorp; Medical Diagnostic Laboratories)

- Comprehensive digestive stool analysis (CDSA) (Genova Diagnostics, Doctor's Data, Diagnostic Solutions [GI-MAP])

As noted above, you can also get specialized tests (shown below) to help evaluate function in each of the Super Seven. However, my recommendation is to begin by following the Clean Genes Protocol—first the Soak and Scrub, then the Spot Cleaning—rather than relying on tests as your first step. Remember, these tests are highly problematic unless you have a skilled health professional to help you evaluate them. As we've seen, the results come back with indications of whether you're "within normal range," but those indications can be extremely misleading.

MTHFR

- **Check for folate receptor antibodies.** If you want to know whether you have antibodies to your folate receptor, this is the test. It can offer a good baseline check that helps you monitor treatment. If you do have antibodies, your remedy is to heal your leaky gut; stop taking folic acid; stop consuming cow's milk dairy products—even in tiny amounts that might be hidden in other foods, like an omelet or a baked good; consume natural folates in your food and supplements; and calm your immune system. (Iliad Neurosciences)

- **Request a fasting test for serum homocysteine.** Have a normal dinner and then have this test the next morning before you have breakfast. Then follow up a month or so later, being sure to eat the same type of dinner and have the blood drawn around the same time of day the next morning. This way you get a more accurate comparison. (Quest Diagnostics, LabCorp)

- **Measure your serum folate**. As I explained in chapter 5, this test is not very useful because of all the folic acid that you might (often unknowingly) be consuming. However, if your reading is high, you might be suffering from one or more of the following conditions: SIBO, folate receptor antibodies, low B_{12}, and/or a blocked Methylation Cycle. If your reading is low, you need to supplement with active folates such as folinic acid and methylfolate while increasing your intake of natural methylfolate in the form of leafy green veggies. (Quest Diagnostics, LabCorp)

- **Test for unmetabolized folic acid.** This test wasn't available at the time of writing, but I'm pushing hard to have a lab develop it. This would be a *true* folic acid test—one that doesn't confuse (unhealthy) folic acid with (healthy) folate.

- **Schedule a methylation panel.** This test checks homocysteine, cysteine, methionine, SAMe, SAH, and SAM:SAH ratio. This provides a useful baseline to see how your Methylation Cycle is doing. It can't tell you *why* it isn't working right, but it definitely shows *whether* it is or isn't. (Doctor's Data)

- **Check for an intrinsic factor deficiency.** If you're consuming vitamin B_{12} yet your levels are still low, you may have antibodies against the stomach cells that absorb this vital nutrient from your diet. Check using the intrinsic factor test. (Specialty Labs, Quest Diagnostics, LabCorp)

COMT

- **Check your estrogen levels.** Use a procedure known as estrogen fractionation to see all three types of estrogen and their components. You can also order the urinary hormone DUTCH test, which is quite accurate for estrogen. The DUTCH test is the easiest and best way to see how your COMT is working. If the catechol estrogens are elevated, that's a sign that your COMT isn't working as well as it should be. (LabCorp, Precision Hormones)

- **Assess urinary neurotransmitters or urinary organic acids to evaluate neurotransmitter breakdown.** If homovanillic acid (HVA) is low, that can be a sign of low dopamine production or slower breakdown. (Great Plains Laboratory, Genova Diagnostics, Doctor's Data, Neuroscience)

- **Measure your tyrosine levels.** If tyrosine is high, it could be because you're eating a lot of protein or you're supplementing with tyrosine. If your tyrosine levels are high and you're feeling anxious, reducing tyrosine-containing supplements or reducing protein intake a bit can help immensely. (Protein intake should be around one gram of protein per two pounds of body weight per day.) If your tyrosine levels are low, it could be because you aren't eating enough protein or you aren't absorbing your protein. You need to support your digestion if you're eating plenty of protein yet your tyrosine is low. (Doctor's Data, LabCorp, Quest Diagnostics)

- **Screen for endocrine-disrupting chemicals via a test that measures glyphosate, DDT, phthalates, and other environmental chemicals.** A toxic chemical profile (for example, GPL-TOX) can help you determine how much effort you need to expend toward reducing your "body burden" of endocrine disrupters. (Great Plains Laboratory)

- **Measure your intracellular RBC magnesium levels.** Magnesium deficiency is common, so you ought to check your intracellular magnesium levels. You can't get your magnesium levels up without taurine, so if your lab results show a low reading, supplement with both compounds. (Quest Diagnostics, LabCorp, Specialty Labs)

DAO

Lab testing for histamine levels is challenging, since the life of histamine is only one minute. Rather than relying on labs to determine what's going on with your DAO, you'd be better off avoiding histamine-containing foods for a few days and noting whether you improve. Then recheck by eating some histamine-containing foods and observing whether your symptoms return. Checking the amount of DAO enzyme via lab testing simply isn't reliable, according to research.

However, here are some related labs that you might consider:

- **Measure urinary histamine.** This is a decent marker of your overall histamine status, because it checks the histamine levels of your stomach content. If elevated, that could be a sign of food allergies or infection. (Quest Diagnostics, LabCorp, Specialty Labs)

- **Measure plasma histamine.** This lab test isn't the best marker, because within minutes after you've consumed high-histamine foods, your blood levels of histamine can return to normal. If this test shows that your plasma histamine level is elevated, that's useful information. If not—and if you believe you have a histamine issue—you might need to redo the test within half an hour of eating. (Quest Diagnostics, LabCorp)

- **Get a comprehensive digestive stool analysis (CDSA).** This test will help you detect the presence of pathogenic bacteria that increase histamine. If such pathogens are found at high levels, you'll want to rebalance your microbiome by decreasing their presence while replenishing other types of bacteria through specific probiotics. (Doctor's Data, Genova Diagnostics, Diagnostic Solutions [GI-MAP])

- **Identify food allergies.** There are two types of immune responses, IgE and IgG. Allergies that trigger IgE responses tend to cause serious issues such as anaphylaxis, so if you have IgE responses, you probably already know that. You're more likely to want to test for IgG responses but testing for both is useful information. (US BioTek)

MAOA

- **Have a urinary organic acid test showing your 5-HIA.** If your 5-HIA levels are high, then you're burning through your serotonin too quickly. If your 5-HIA levels are low, your body may not be breaking down your serotonin well, or you may have low levels of serotonin building blocks such as tryptophan and vitamin B_6. (Great Plains Laboratory, Genova Diagnostics)

- **Measure your tryptophan levels.** If your urinary or blood levels of tryptophan are high, that might mean you're consuming a lot of carbohydrates or that you're not able to turn tryptophan into serotonin. This inability can be caused by a slow MAOA. (Quest Diagnostics, LabCorp, Great Plains Laboratory, Genova Diagnostics)

- **Evaluate your vitamin B_6.** If these levels are low, your ability to make serotonin is reduced, dirtying your MAOA. Inadequate vitamin B_6 is one factor that leads to increased concentrations of xanthurenate and kynurenate in urine. You can measure these compounds—and thereby infer your B_6 status—through urinary organic acid testing. (Quest Diagnostics, LabCorp, Great Plains Laboratory, Genova Diagnostics)

- **Evaluate your vitamin B_2.** If your vitamin B_2 levels are low, your ability to support your MAOA is reduced and that gene may become slow as a result. If you don't have sufficient riboflavin, compounds such as adipate, suberate, and ethylmalonate may increase in your urine, demonstrating that deficiency. (Quest Diagnostics, LabCorp, Great Plains Laboratory, Genova Diagnostics)

Be aware that inflammation or infection might be one reason why you have low levels of 5-HIA, tryptophan, B_6, and/or B_2. Tryptophan can also move through another enzyme called IDO1 (in addition to MAOA), which increases during times of stress, infection, and inflammation. Those three factors—stress, infection, and inflammation—use up your tryptophan, making it appear that you have a faster MAOA gene, whereas in reality your MAOA isn't able to function well because of the shortage of tryptophan.

The way to check this is by running a urinary organic acid test (see advice on B$_6$ testing, above). Look for elevated levels of quinolinate and kynurenate. (Great Plains Laboratory)

GST/GPX

Glutathione levels are measured by your health professional in order to understand how well your body is handling free radicals and to evaluate the overall state of your antioxidant potential. Basically, the higher your glutathione levels, the healthier you are, while lower glutathione levels correlate to ill health.

- **Measure RBC glutathione peroxidase.** This marker demonstrates how well the GST is working based on levels of xenobiotics and/or hydrogen peroxide. This can be an expensive test and is difficult to find. (Genova Diagnostics)

- **Evaluate lipid peroxidation.** Test results demonstrate the degree of damage to cell membranes. (Quest Diagnostics, LabCorp)

- **Measure RBC glutathione.** This will help determine levels of glutathione in your red blood cells. (Doctor's Data, Genova Diagnostics)

- **Evaluate urinary organic acid markers for riboflavin deficiency.** This result tells you whether you're able to recycle your glutathione. Elevated levels of any of the following acids denote a possible riboflavin deficiency: succinic acid, fumaric acid, 2-oxoglutaric acid, or glutaric acid. (Quest Diagnostics, LabCorp, Genova Diagnostics, Great Plains Laboratory)

- **Evaluate selenium.** This is done via a blood test. Too much selenium is toxic, and too little means you lack a key cofactor, so yet again, you need balance. I've seen people whose selenium levels rose too high after they got an intravenous infusion of various nutrients, including selenium. Make sure that your doctor isn't giving you too much—and that you aren't taking too much via supplements. (Quest Diagnostics, LabCorp)

NOS3

If you have a history of cardiovascular issues in your family, or if there are indications that you have a dirty NOS3, keeping an eye on your labs is important:

- You want your homocysteine to measure around 7.

- Your lipid peroxides should be low.

- You need your Lp(a) (an inflammatory type of cholesterol) in normal range, as well as your hs-CRP. (Quest Diagnostics, LabCorp)

Checking for bacterial, viral, and mold infections is also key, because any infection will use up your arginine and increase your cardiovascular risk. As you may remember, you need arginine to help support your NOS3.

Here are some labs to consider:

- **Measure blood amino acids.** This checks your levels of arginine, ornithine, and citrulline so that you know whether your NOS3 has the nutrients it needs to function. (Quest Diagnostics, LabCorp, Doctor's Data, Genova Diagnostics, Great Plains Laboratory)

- **Evaluate your ADMA.** This can be an expensive test, but if your ADMA levels are elevated, that shows clearly that your NOS3 isn't working well. (Genova Diagnostics, Mayo Clinic, Cleveland Heart Lab)

- **Measure your homocysteine.** If your homocysteine levels are elevated, you can assume that your NOS3 isn't working well. (Quest Diagnostics, LabCorp)

- **Measure your lipid peroxides.** Again, if these levels are elevated, you can assume that your NOS3 isn't working well. (Quest Diagnostics, LabCorp)

- **Consider a comprehensive digestive stool analysis (CDSA).** This test evaluates your microbiome for the following bacteria: *Streptococcus (or Enterococcus) faecalis, Mycoplasma, Bacillus, Pseudomonas aeruginosa, Halobacterium, Spirochaeta,* and possibly *Clostridium.* If present, these are consuming your arginine, causing a shortage that might be harming your NOS3. (Diagnostic Solutions [GI-MAP], Genova Diagnostics, Doctor's Data)

- **Evaluate fasting insulin.** If these levels are elevated, then your NOS3 might be having to work extra hard. As a result, instead of making nitric oxide (good), it might be making superoxide (bad). (Quest Diagnostics, LabCorp)

- **Assess serum nitrite and serum nitrate.** Checks your levels of nitrites and nitrates. These can be high or low during times of inflammation, infection, or cardiovascular issues, so assessing them can be useful. (Quest Diagnostics)

- **Measure estrogen levels.** If your estrogens are low, as assessed by the DUTCH test, your NOS3 may not be working well and you'll need to figure out how to support it. If they're elevated, your NOS3 might be working too hard; in that case, you need to reduce them. (Precision Hormones)

- **Schedule a sleep study.** If you snore or are constantly tired, consider a sleep study. Evaluating how you sleep and breathe at night could save your life. Sleep apnea is common. The causes of this disorder are many, but first you need to suss out whether you have it. A good place to start is at-home sleep testing, which isn't as thorough as in-office testing, but not nearly as expensive either. (NovaSom, for home test kits)

PEMT

- **Get a serum choline test.** If your serum choline levels are low, you know that your PEMT is stressed by working hard to produce choline. (Quest Diagnostics, LabCorp)

- **Measure creatine phosphokinase (CPK).** This compound is elevated when you're deficient in phosphatidylcholine, so measuring CPK is a useful way to evaluate muscle membrane damage and potential injury to muscles, heart, or brain. (Quest Diagnostics, LabCorp)

- **Evaluate DHEA-S.** This compound is commonly low; when it is, that deficiency contributes to muscle weakness. (Quest Diagnostics, LabCorp, Precision Hormones [DUTCH test])

- **Measure your ALT.** This is a liver enzyme. Elevated, it demonstrates that the phosphatidylcholine levels need to increase. (Quest Diagnostics, LabCorp)

- **Evaluate lipid peroxides.** If these are elevated, then you know that cell membranes are being damaged and your body needs more phosphatidylcholine. (Quest Diagnostics, LabCorp)

- **Measure TMAO.** If these levels are elevated, it may be due to choline or phosphatidylcholine supplementation. Order a comprehensive digestive and stool analysis (CDSA) to determine what's going on. It's important to avoid dairy products if these levels are high. Higher TMAO levels are associated with poor metabolic control (potentially leading to diabetes) and with kidney issues. (Cleveland Heart Lab)

- **Measure GGT.** This is an early marker of fatty liver. (Quest Diagnostics, LabCorp)

- **Use the fatty liver index calculator.** This tool has been developed to help identify fatty liver early—a great aid for you and your health professional. (Quest Diagnostics, LabCorp)

- **Take the SIBO breath test.** This can help determine whether you have SIBO. (Commonwealth Laboratories)

- **Measure your fasting insulin.** This is a great way to see how well your metabolism is doing. If your fasting insulin is elevated, significant lifestyle, environmental, and dietary changes are needed.

- **Measure your LDL and HDL cholesterol, as well as triglycerides, via an advanced cholesterol panel (such as VAP).** People who are deficient in choline typically also show reduced blood concentrations of LDL cholesterol. Further indications of PEMT trouble include low HDL and high triglycerides. (Quest Diagnostics)

- **Measure estrogen.** Because your PEMT is stimulated by estrogen, low estrogen levels will slow down PEMT function—unless you have a SNP that causes your dirty PEMT not to react to estrogen. High estrogen uses up SAMe, meaning that your body becomes less able to make phosphatidylcholine. (Precision Hormones [DUTCH test])

- **Measure homocysteine levels.** Homocysteine levels higher than 7 could indicate Methylation Cycle issues, affecting the level of SAMe and the production of phosphatidylcholine. (Quest Diagnostics, LabCorp)

- **Evaluate the SAM:SAH ratio as well as SAH levels.** These give you an indication of how your Methylation Cycle is functioning and whether your PEMT has been adversely affected. (Doctor's Data)

- **Measure folate and B$_{12}$ levels.** If you're deficient in either of these vitamins, your Methylation Cycle won't function well and your PEMT will be affected. (Quest Diagnostics, LabCorp)

- **Identify bacterial infection by assessing LPS.** Elevated levels of LPS reveal the presence of bacterial infection, which can affect your Methylation Cycle and therefore your PEMT function. (Medical Diagnostic Laboratories, DNA Connexions, Quest Diagnostics, LabCorp, Specialty Labs)

- **Test for viral infections, particularly hepatitis (A, B, and C), Coxsackie, and Epstein-Barr.** These increase oxidative stress and inflammation, which affect your Methylation Cycle and therefore your PEMT function. (Medical Diagnostic Laboratories, DNA Connexions, Quest Diagnostics, LabCorp, Specialty Labs)

Laboratories

Cleveland Heart Lab: http://www.clevelandheartlab.com

Commonwealth Laboratories: http://commlabsllc.com

Diagnostic Solutions: https://diagnosticsolutionslab.com

Direct Labs: http://www.directlabs.com

DNA Connexions: http://dnaconnexions.com

Doctor's Data: https://www.doctorsdata.com

Dr. Lal PathLabs: https://www.lalpathlabs.com

Genova Diagnostics: https://www.gdx.net

Great Plains Laboratory: https://www.greatplainslaboratory.com

Iliad Neurosciences: http://iliadneuro.com

LabCorp: https://www.labcorp.com

Mayo Clinic: http://www.mayoclinic.org

Medical Diagnostic Laboratories: http://www.mdlab.com

Precision Hormones (DUTCH test): https://dutchtest.com

Specialty Labs: http://www.specialtylabs.com

Quest Diagnostics: http://www.questdiagnostics.com/home.html

US BioTek: http://www.usbiotek.com

Genetic Testing and Evaluation

If you're interested in finding out more about your ancestry, you have a wide range of options. But if your focus is on genetics as it relates to *health*, the companies listed in this appendix are your best bet. They offer either genetic testing or help evaluating the results.

Testing

Let's start with the testing options:

- **Genos Research** (https://genos.co). As of April 2017, this company tests fifty times more of your DNA than 23andMe. They also give you access to your raw data. Overall, the value is fantastic. However, they don't test the *regulatory regions* of your DNA: the genes that control how other genes are turned on or off. Instead, they test your entire exome, which lies within your regulatory regions. This is important to realize going in, because some genes—PEMT, for example—have SNPs you'd want to know about in the regulatory regions.

- **23andMe** (https://www.23andme.com). This company provides two testing options: with a health report and without. The health report is useful if you want their advice on what the data means. However, you can pay less to get just your data and then use a genetic evaluation tool (see below).

- **Courtagen** (http://www.courtagen.com). This company offers specialty panels for various conditions, such as autism, seizure disorder, or mitochondrial disease. Insurance may cover this testing.

- **GeneSight** (https://genesight.com). GeneSight offers a panel that's useful if you're not responding well to psychiatric medications. Insurance may cover it.

- **Arivale** (https://www.arivale.com). This company offers comprehensive genetic and lab testing, with health coaches to guide you. It's expensive, but you do seem to receive a comprehensive service, rather than being given your lab results with little guidance on how to use them.

- **Pathway Genomics** (https://www.pathway.com). They offer a number of different genetic testing options, including corporate wellness programs. Insurance may cover their lab testing.

- **DNAFit** (https://www.dnafit.com). This company offers tests tailored to fitness, sports performance, and general well-being.

- **uBiome** (https://ubiome.com). This company evaluates the DNA of your microbiome, which is pretty fascinating, given that the genes of your microbiome outnumber your human genes by a factor of 150 to 1 and have an enormous impact on your health. This company can also specifically check the biome of your throat, ears, nose, throat, and skin.

Evaluation

As I explained in chapter 1, genetic test results often lead to massive confusion. You receive all kinds of information with either no recommendations or conflicting recommendations: "Take a lot of vitamin XYZ to respond to SNP A; avoid vitamin XYZ to respond to SNP B; take a moderate amount of vitamin XYZ to respond to SNP C." What are you supposed to do with that?

The answer may lie with one of the new companies developed to help you evaluate your test results and turn them into a specific, actionable plan. Here are three companies—one of which is my own—that have sprung up to respond to this need:

- **StrateGene** (www.strategene.org). This is the company I developed and continue to run. We provide an integrated approach to clinically relevant SNPs using graphical illustrations. As you learned in this book, you need to understand not only which SNPs you have but also how these genes are affected by your lifestyle, diet, environment, and nutrients. StrateGene offers this information. Your purchase includes access to a private Facebook group for ongoing community and support.

- **Opus23** (https://www.datapunk.net/opus23). This company is available to health professionals only. It was developed by Dr. Peter D'Adamo, a brilliant naturopath and the author of *Eat Right 4 Your Type*. Opus23 offers a powerful suite of tools to dig deep into a patient's raw data from uBiome or 23andMe. Consider recommending it to your practitioner.

- **Promethease** (https://promethease.com). This online DNA reporting tool uses your raw genetic data to evaluate your SNPs. It provides a lot of SNP information but doesn't discuss how the genes are affected by your lifestyle, diet, or environment. The results can be overwhelming; they're very much about predicting your chances of disease rather than giving you actionable information for health. I would recommend using this tool in addition to StrateGene—but only when you're emotionally prepared for this type of information.

Mold and Indoor Air Quality Testing

Mold is a massively important issue that more people need to be aware of. If you're chronically ill and not getting better, mold very likely plays a part in your illness. Please evaluate your home, car, office, boat—and anywhere else you spend time—for mold.

I had a patient with chronic congestion that simply wouldn't go away. She was a teacher, so eventually I had her call in inspectors to evaluate her school. Turns out the building was so contaminated by mold that they had to demolish it! I could have just treated her congestion, but as it turned out, I helped thousands of people. Please, check for mold— and also for a number of other common indoor air pollutants, including radon, carbon monoxide, dust mites, and formaldehyde (to name just a few).

A good starting point is often an at-home mold test kit, available at your local hardware store and online (see below for one recommenda- tion). If this doesn't work, call in the professionals to make an evalua- tion. Once you identify the mold, you need to have a professional come in and remediate it.

Here are some useful resources related to mold and indoor air quality:

- **DIY Mold Test.** This is an easy-to-use test kit that you can use initially to evaluate whether you have mold. It's widely available (e.g., hardware stores and www.amazon.com), and it comes with an expert phone consultation.

- **The American Lung Association** (www.lung.org). This association is a great resource for learning about potential problems and solutions for your indoor air.

- **Indoor Air Quality Association** (http://www.iaqa.org/find-a-pro). This all-inclusive organization focuses on air quality and solving indoor environmental problems, including issues around construction and remodeling, research, school contamination, storm damage, and mold.

- **National Association of Mold Remediators and Inspectors** (https://www.namri.org/index.php). Whether you're searching for a reputable mold-removal company, or seeking knowledge about mold-removal services generally, the National Association of Mold Remediators and Inspectors provides essential information for your residential or commercial property.

Notes

Introduction: Your Genes Are Not Your Destiny!

2 *[Diet] reshaped their genetic destiny:* Wolff, G. L., et al., "Maternal epigenetics and methyl supplements affect agouti gene expression in Avy/a mice," *FASEB Journal*, August 1998, http://www.fasebj.org/content/12/11/949.abstract.

1. Cleaning Up Your Dirty Genes

18 *[Air, water, food] … 129 million industrial chemicals:* "CAS Registry: The gold standard for chemical substance information," *CAS: A Division of the American Chemical Society*, accessed April 2017, http://www.cas.org/content/chemical-substances.

21 *About twenty thousand genes:* Ezkurdia, I., et al., "Multiple evidence strands suggest that there may be as few as 19,000 human protein-coding genes," *Human Molecular Genetics*, 16 June 2014, https://www.ncbi.nlm.nih.gov/pmc/articles/PMC4204768.

21 *More than ten million known genetic polymorphisms (SNPs):* "Genetics home reference: Your guide to understanding genetic conditions," *US National Library of Medicine*, 4 April 2017, https://ghr.nlm.nih.gov/primer/genomicresearch/snp.

23 *Decreased risk of colon cancer:* Xie, S. Z., et al., "Association between the MTHFR C677T polymorphism and risk of cancer: Evidence from 446 case-control studies," *Tumour Biology*, 17 June 2015, https://www.ncbi.nlm.nih.gov/pubmed/26081619.

24 *Increased risk of stomach cancer:* Xie, S. Z., et al., "Association between the MTHFR C677T polymorphism and risk of cancer: Evidence from 446 case-control studies," *Tumour Biology*, 17 June 2015, https://www.ncbi.nlm.nih.gov/pubmed/26081619.

24 *Serious long-term problems:* Wood, J. D., "Histamine, mast cells, and the enteric nervous system in irritable bowel syndrome, enteritis, and food allergies," *Gut*, April 2006, https://www.ncbi.nlm.nih.gov/pmc/articles/PMC1856149.

29 *Born with over two hundred chemicals:* "Body burden: The pollution of newborns," *Environmental Working Group,* 14 July 2005, http://www.ewg.org/research/body-burden -pollution-newborns.

29 *Avoid the worst offenders:* "EWG's 2017 shopper's guide to pesticides in produce," *Environmental Working Group,* April 2017, https://www.ewg.org/foodnews/summary.php.

34 *Incredibly focused and determined:* Tsai, A. J., et al., "Heterozygote advantage of the MTHFR C677T polymorphism on specific cognitive performance in elderly Chinese males without dementia," *Dementia and Geriatric Cognitive Disorders,* 13 October 2011, https://www.ncbi.nlm.nih.gov/pubmed/21997345.

2. Gene Secrets: What They Didn't Teach You in Science Class

41 *Genes that contribute to cancer:* Tost, J., "DNA methylation: An introduction to the biology and the disease-associated changes of a promising biomarker," *Molecular Biotechnology,* January 2010, https://www.ncbi.nlm.nih.gov/pubmed/19842073.

42 *Burning fat instead of storing it:* Podlepa, E. M., Gessler, N. N, and Bykhovski, Via, "The effect of methylation on the carnitine synthesis," *Prikladaia Biokhimiia i Mikrobiolgiia,* March–April 1990, https://www.ncbi.nlm.nih.gov/pubmed/2367349.

42 *Burn fuel as efficiently as possible:* Wenyi, X. U., et al., "Epigenetics and cellular metabolism," *Genetics and Epigenetics,* 25 September 2016, https://www.ncbi.nlm.nih .gov/pmc/articles/PMC5038610; Donohoe, D. R., Bultman, S. J., "Metaboloepigenetics: Interrelationships between energy metabolism and epigenetic control of gene expression," *Journal of Cell Physiology,* September 2012, https://www.ncbi.nlm.nih.gov/pmc/articles /PMC3338882.

43 *The 2.5 million that die every second:* "How many cells do we have in our body?" *UCSB Science Line,* 2015, http://scienceline.ucsb.edu/getkey.php?key=3926.

43 *Pain, fatigue, inflammation, and fatty liver:* Sanders, L. M., and Zeisel, S. H., "Choline," *Nutrition Today,* 2007, https://www.ncbi.nlm.nih.gov/pmc/articles/PMC2518394.

43 *[Nausea, vomiting, or gallbladder issues] … poor methylation:* Jarnfelt-Samsioe, A. "Nausea and vomiting in pregnancy: A review," *Obstetrical & Gynecological Survey,* July 1987, https://www.ncbi.nlm.nih.gov/pubmed/3614796; Pusi, T., and Beuers, U., "Intrahepatic cholestasis of pregnancy," *Orphanet Journal of Rare Disease,* 29 May 2007, https://www .ncbi.nlm.nih.gov/pmc/articles/PMC1891276.

43 *[Birth defects] results of a methylation deficiency:* Blom, H. J., et al., "Neural tube defects and folate: Case far from closed," *Nature Reviews Neuroscience,* September 2006, http:// pubmedcentralcanada.ca/pmcc/articles/PMC2970514; Imbard, A., Benoist, J.-F., Blom, H. J., "Neural tube defects, folic acid and methylation," *International Journal of Environmental Research and Public Health,* 17 September 2013, https://www.ncbi.nlm .nih.gov/pmc/articles/PMC3799525.

43 *Methylation also produces creatine:* Bronsan, J. T., Da Silva, R. P., and Bronsan, M. E., "The metabolic burden of creatine synthesis," *Amino Acids,* May 2011, https://www.ncbi .nlm.nih.gov/pubmed/21387089.

43 *Muscular aches and pains:* Onodi, L., et al., "Creatine treatment to relieve muscle pain caused by thyroxine replacement therapy," *Pain Medicine,* 12 April 2012, https:// academic.oup.com/painmedicine/article-lookup/doi/10.1111/j.1526–4637.2012.01354.x.

45 *To clear harmful chemicals and excess hormones:* Dawling, S., et al., "Catechol-O-methyltransferase (COMT)-mediated metabolism of catechol estrogens: Comparison of wild-type and variant COMT isoforms," *Cancer Research*, 15 September 2001, https://www.ncbi.nlm.nih.gov/pubmed/11559542.

45 *Methylation also affects your ability:* Prudova, A., et al., "S-adenosylmethionine stabilizes cystathionine ß-synthase and modulates redox capacity," *Proceedings of the National Academy of Sciences*, 9 March 2006, http://www.pnas.org/content/103/17/6489.full.

46 *Methylation helps your immune system:* Lei, W., et al., "Abnormal DNA methylation in CD4+ T cells from patients with systemic lupus erythematosus, systemic sclerosis, and dermatomyositis," *Scandinavian Journal of Rheumatology*, 2009, https://www.ncbi.nlm.nih.gov/pubmed/19444718.

46 *Atherosclerosis (hardening of the arteries) and hypertension:* Zhong, J., Agha, G., and Baccarelli, A. A., "The role of DNA methylation in cardiovascular risk and disease," *Circulation Research*, 8 January 2016, http://circres.ahajournals.org/content/118/1/119.

46 *Helps prevent DNA errors:* Bluont, B. C., et al., "Folate deficiency causes uracil misincorporation into human DNA and chromosome breakage: Implications for cancer and neuronal damage," *Proceedings of the National Academy of Sciences*, 1 April 1997, https://www.ncbi.nlm.nih.gov/pubmed/9096386.

48 *Folic acid is unnatural:* Bailey, S. W., and Ayling, J. E., "The extremely slow and variable activity of dihydrofolate reductase in human liver and its implications for high folic acid intake," *Proceedings of the National Academy of Sciences*, 22 July 2009, http://www.pnas.org/content/106/36/15424.long.

48 *Folic acid blocks methylation:* Christensen, K. E., et al., "High folic acid consumption leads to pseudo-MTHFR deficiency, altered lipid metabolism, and liver injury in mice," *American Journal of Clinical Nutrition*, March 2015, https://www.ncbi.nlm.nih.gov/pubmed/25733650.

48 *Take folate [instead of folic acid]:* Lynch, B., "Folic acid and pregnancy: Is folic acid the right choice?" *YouTube*, 7 September 2016, https://www.youtube.com/watch?v=tnVRv0zGsFY&t=603s.

48 *Requiring U.S. manufacturers to "enrich" the following foods:* "Folate," *National Institutes of Health*, accessed April 2017, https://ods.od.nih.gov/factsheets/Folate-Health Professional.

49 *The wrong amount of exercise:* Reynolds, G., "How exercise changes our DNA," *Well*, 17 December 2014, https://well.blogs.nytimes.com/2014/12/17/how-exercise-changes-our-dna/?_r=1.

49 *Poor sleep:* Kirkpatrick, B., "The Epigenetics of sleep: 3 reasons to catch more zzz's," *What Is Epigenetics*, 3 March 2015, http://www.whatisepigenetics.com/the-epigenetics-of-sleep-3-reasons-to-catch-more-zzzs.

49 *When your body is under stress:* Bing, Y., et al., "Glucocorticoid-induced S-adenosylmethionine enhances the interferon signaling pathway by restoring STAT1 protein methylation in hepatitis B virus-infected cells," *Journal of Biological Chemistry*, 30 September 2014, http://www.jbc.org/content/289/47/32639.full.

3. What's Your Genetic Profile?

55 *Affects estrogen metabolism:* Cussenot, O., "Combination of polymorphisms from genes related to estrogen metabolism and risk of prostate cancers: The hidden face of estrogens," *Journal of Clinical Oncology,* August 2007, http://ascopubs.org/doi/full/10 .1200/JCO.2007.11.0908.

5. MTHFR: Methylation Master

76 *Decreased risk of colon cancer:* Xie, S. Z., et al., "Association between the MTHFR C677T polymorphism and risk of cancer: Evidence from 446 case-control studies," *Tumour Biology,* 17 June 2015, https://www.ncbi.nlm.nih.gov/pubmed/26081619.

77 *More than one hundred SNPs:* "MTHFR[all]," *National Center for Biotechnology Information, U.S. National Library of Medicine,* accessed April 2017, https://www.ncbi .nlm.nih.gov/clinvar/?term=MTHFR[all].

77 *[Italian diets] support healthy methylation:* Wilcken, W., et al., "Geographical and ethnic variation of the 677C>T allele of 5,10 methylenetetrahydrofolate reductase (MTHFR): Findings from over 7000 newborns from 16 areas worldwide," *Journal of Medical Genetics,* 2003, http://jmg.bmj.com/content/40/8/619.

78 *Disorders that researchers have associated with MTHFR SNPs:* "Genopedia: MTHFR," *Center for Disease Control and Prevention,* accessed April 2017, https://phgkb.cdc.gov /HuGENavigator/huGEPedia.do?firstQuery=MTHFR&geneID=4524&typeSubmit =GO&check=y&typeOption=gene&which=2&pubOrderType=pubD.

86 *Cobalamin/B$_{12}$ [foods]:* "Vitamin B$_{12}$," *Oregon State University Linus Pauling Institute's Micronutrient Information Center,* accessed April 2017, http://lpi.oregonstate.edu/mic /vitamins/vitamin-B12#food-sources.

6. COMT: Focus and Buoyancy, or Mellowness and Calm

95 *Catechols are compounds found in:* "COMT gene," *Genetic Home Reference,* 11 April 2017, https://ghr.nlm.nih.gov/gene/COMT#resources.

106 *Methylphenidate may increase:* Miyazak, I., and Asanuma, M. "Approaches to prevent dopamine quinone-induced neurotoxicity," *Neurochemical Research,* 4 September 2008, http://link.springer.com/article/10.1007/s11064-008-9843-1; Sadasivan, S., et al., "Methylphenidate exposure induces dopamine neuron loss and activation of microglia in the basal ganglia of mice," *PLOS One,* 21 March 2012, http://journals.plos.org/plosone /article?id=10.1371/journal.pone.0033693; Espay, A. J., et al., "Methylphenidate for gait impairment in Parkinson disease," *American Academy of Neurology,* 5 April 2011, https:// www.ncbi.nlm.nih.gov/pmc/articles/PMC3068005.

106 *Adderall can also generate dopamine quinone:* German, C. L., Hanson, G. R., and Fleckenstein, A. E., "Amphetamine and methamphetamine reduce striatal dopamine transporter function without concurrent dopamine transporter relocalization," *Journal of Neurochemistry,* 23 August 2012, https://www.ncbi.nlm.nih.gov/pmc/articles/PMC3962019.

109 *Two common reasons for magnesium deficiency:* Janett, S., et al., "Hypomagnesemia induced by long-term treatment with proton-pump inhibitors," *Gastroenterology Research & Practice,* 4 May 2015, https://www.ncbi.nlm.nih.gov/pubmed/26064102; Kynast-Gales,

S. A., and Massey, L. K., "Effect of caffeine on circadian excretion of urinary calcium and magnesium," *Journal of American College of Nutrition*, October 1994, https://www.ncbi.nlm.nih.gov/pubmed/7836625.

112 *[Damage from] dioxins:* Liu, J., et al., "Variants in maternal COMT and MTHFR genes and risk of neural tube defects in offspring," *Metabolic Brain Disease*, 4 July 2014, https://www.ncbi.nlm.nih.gov/pubmed/24990354.

113 *Hugs raise dopamine:* "The power of love: Hugs and cuddles have long-term effects," *NIH News in Health*, February 2007, https://newsinhealth.nih.gov/2007/february/docs/01 features_01.htm.

7. DAO: Oversensitivity to Foods

119 *DAO enzyme, which is found in most organs:* "AOC1 gene (protein coding)," *Gene Cards*, accessed April 2017, http://www.genecards.org/cgi-bin/carddisp.pl?gene=AOC1# expression.

120 *Ways to track down this dirty gene:* "AOC1 gene (protein coding)," *Gene Cards*, accessed April 2017, http://www.genecards.org/cgi-bin/carddisp.pl?gene=AOC1#expression.

122 *[DAO causing] irritable bowel disorders:* Xie, H., He, S.-H., "Roles of histamine and its receptors in allergic and inflammatory bowel diseases," *World Journal of Gastroenterology*, 21 May 2005, https://www.ncbi.nlm.nih.gov/pmc/articles/PMC4305649.

129 *Copper … turnip greens:* "Copper," *Oregon State University Linus Pauling Institute's Micronutrient Information Center*, accessed April 2017, http://lpi.oregonstate.edu/mic /minerals/copper.

8. MAOA: Mood Swings and Carb Cravings

135 *Neurotransmitters will be more stable:* Fernstrom, J. D., et al., "Diurnal variations in plasma concentrations of tryptophan, tryosine, and other neutral amino acids: Effect of dietary protein intake," *American Journal of Clinical Nutrition*, September 1979, https://www.ncbi .nlm.nih.gov/pubmed/573061.

140 *MAOA is involved in processing neurotransmitters:* "Genopedia: MAOA," *Center for Disease Control and Prevention*, accessed April 2017, https://phgkb.cdc.gov/HuGENavigator /huGEPedia.do?firstQuery=MAOA&geneID=4128&typeSubmit=GO&check=y&type Option=gene&which=2&pubOrderType=pubD.

141 *[Hydrogen peroxide can lead to] neurological problems:* Balmus, I. M., et al., "Oxidative stress implications in the affective disorders: Main biomarkers, animal models relevance, genetic perspectives, and antioxidant approaches," *Oxidative Medicine and Cellular Longevity*, 1 August 2016, https://www.ncbi.nlm.nih.gov/pubmed/27563374.

144 *Tryptophan … asparagus:* "Foods highest in tryptophan," *Self Nutrition Data*, accessed April 2017, http://nutritiondata.self.com/foods-011079000000000000000.html?maxCount=60.

9. GST/GPX: Detox Dilemmas

152 *[Dirty GST related to] increased inflammation:* Luo, L., et al., "Recombinant protein glutathione S-transferases P1 attenuates inflammation in mice," *Molecular Immunology*,

28 October 2008, https://www.ncbi.nlm.nih.gov/pubmed?cmd=search&term=18962 899&dopt=b.

152 *[Dirty GST related to] overweight/obesity:* Chielle, E. O., et al. "Impact of the Ile105Val polymorphism of the glutathione S-transferase P1 (GSTP1) gene on obesity and markers of cardiometabolic risk in young adult population." *Experimental and Clinical Endocrinol and Diabetes.* 2017 May;125(5):335–341. https://www.ncbi.nlm.nih.gov /pubmed/27657993.

153 *Discolor and damage your hair:* Wood, J. M., et al., "Senile hair graying: H2O2-mediated oxidative stress affects human hair color by blunting methionine sulfoxide repair," *FASEB Journal,* 23 February 2009, https://www.ncbi.nlm.nih.gov/pubmed /19237503.

155 *Many types of GST gene:* "GST," *Gene Cards,* accessed April 2017, http://www.genecards .org/Search/Keyword?queryString=%22GST%22.

156 *Protecting you against chemical and oxidative stress:* Ziglari, T., and Allameh, A., "The significance of glutathione conjugation in aflatoxin metabolism," *Aflatoxins—Recent Advances and Future Prospects,* 23 January 2013, https://www.intechopen.com/books /aflatoxins-recent-advances-and-future-prospects/the-significance-of-glutathione -conjugation-in-aflatoxin-metabolism.

158 *Easier to achieve your optimal weight:* Crinnion, W., "Clean, green, and lean: Get rid of the toxins that make you fat," *Amazon,* accessed April 2017, https://www.amazon.com /Clean-Green-Lean-Toxins-That-ebook/dp/B00DNKYI8E/ref=tmm_kin_swatch_0? _encoding=UTF8&qid=&sr=.

159 *When glutathione levels drop:* Kut, J. L., et al., "Regulation of murine T-lymphocyte function by spleen cell-derived and exogenous serotonin," *Immunopharmacology & Immunotoxicology,* 1992, https://www.ncbi.nlm.nih.gov/pubmed/1294623.

159 *Damaged glutathione ... contributes to further damage:* Mulherin, D. M., Thurnham, D. I., and Situnayake, R. D., "Glutathione reductase activity, riboflavin status, and disease activity in rheumatoid arthritis," *Annals of the Rheumatic Diseases,* November 1996, https://www.ncbi.nlm.nih.gov/pubmed/8976642; Taniguchi, M., and Hara, T., "Effects of riboflavin and selenium deficiencies on glutathione and its relating enzyme activities with respect to lipid peroxide content of rat livers," *Journal of Nutritional Science and Vitaminology,* June 1983, https://www.ncbi.nlm.nih.gov /pubmed/6619991.

160 *Selenium ... brazil nuts:* "Selenium," *National Institutes of Health,* accessed April 2017, https://ods.od.nih.gov/factsheets/Selenium-HealthProfessional.

161 *Lungs need adequate hydrogen sulfide:* Wang, P., et al., "Hydrogen sulfide and asthma," *Experimental Physiology,* 10 June 2011, https://www.ncbi.nlm.nih.gov/pubmed/21666034.

163 *[Fiber] binds to xenobiotics:* Stein, K., et al., "Fermented wheat aleurone induces enzymes involved in detoxification of carcinogens and in antioxidative defence in human colon cells," *British Journal of Nutrition,* 28 June 2010, https://www.ncbi.nlm.nih.gov/pubmed/20579402.

164 *Lots of choices [to sweat]:* Genuis, S. J., et al., "Blood, urine, and sweat (BUS) study: Monitoring and elimination of bioaccumulated toxic elements," *Archives of Environmental Contamination and Toxicology,* 6 November 2010, https://www.ncbi.nlm.nih.gov/pubmed /21057782.

10. NOS3: Heart Issues

168 *Ended up having a stroke:* Loscalzo, J., et al., "Nitric oxide insufficiency and arterial thrombosis," *Transactions of the American Clinical and Climatological Association*, 2000, https://www.ncbi.nlm.nih.gov/pmc/articles/PMC2194373/pdf/tacca00005-0216.pdf.

168 *Angiogenesis:* Adair, T. H., and Montani, J. P., "Overview of angiogenesis," *Angiogenesis*, 2010, https://www.ncbi.nlm.nih.gov/books/NBK53238.

168 *If you don't have healthy angiogenesis:* Lee, P. C., et al., "Impaired wound healing and angiogenesis in eNOS-deficient mice," *American Journal of Physiology*, October 1999, https://www.ncbi.nlm.nih.gov/pubmed/10516200; Soneja, A., Drews, M., and Malinski, T., "Role of nitric oxide, nitroxidative and oxidative stress in wound healing," *Pharmacological Reports*, 2005, https://www.ncbi.nlm.nih.gov/pubmed/16415491.

169 *Essential hypertension:* Kivi, R., "Just the essentials of essential hypertension," *Health Line*, 21 December 2015, http://www.healthline.com/health/essential-hypertension#overview1.

169 *At risk for cardiovascular disease.* Guck, T. P., et al., "Assessment and treatment of depression following myocardial infarction," *American Family Physician*, 15 August 2001, http://www.aafp.org/afp/2001/0815/p641.html.

170 *Blood flow and blood vessel formation:* "NOS3," *Gene Cards*, accessed April 2017, http://www.genecards.org/cgi-bin/carddisp.pl?gene=NOS3&keywords=NOS3.

170 *Can lead to blood clots:* Loscalzo, J., et al., "Nitric oxide insufficiency and arterial thrombosis," *Transactions of the American Clinical and Climatological Association*, 2000, https://www.ncbi.nlm.nih.gov/pmc/articles/PMC2194373/pdf/tacca00005-0216.pdf.

170 *Issues result from a dirty NOS3:* Burke, T., "Nitric oxide and its role in health and diabetes," *Diabetes in Control*, accessed April 2017, http://www.diabetesincontrol.com/wp-content/uploads/2015/10/nitric-oxide.pdf.

171 *Diabetic complications are the result:* Giacco, F., and Brownlee, M., "Oxidative stress and diabetic complications," *Circulation Research*, 29 October 2010, https://www.ncbi.nlm.nih.gov/pmc/articles/PMC2996922; Katakam, P. V., et al., "Insulin-induced generation of reactive oxygen species and uncoupling of nitric oxide synthase underlie the cerebrovascular insulin resistance in obese rats," *Journal of Cerebral Blood Flow and Metabolism*, May 2012, https://www.ncbi.nlm.nih.gov/pubmed/22234336.

171 *[Potential for] congenital heart defect:* Feng, Q., et al., "Development of heart failure and congenital septal defects in mice lacking endothelial nitric oxide synthase," *Circulation*, 13 August 2002, https://www.ncbi.nlm.nih.gov/pubmed/12176963; Liu, Y., et al., "Nitric oxide synthase-3 promotes embryonic development of atrioventricular valves," *PLOS One*, 29 October 2013, https://www.ncbi.nlm.nih.gov/pubmed/24204893.

171 *Most common birth defect in humans:* Liu, Y., and Feng, Q. "NOing the heart: Role of nitric oxide synthase-3 in heart development," *Differentiation*, July 2012, https://www.ncbi.nlm.nih.gov/pubmed/22579300.

172 *[Dirty NOS3] contributing to more than four hundred conditions:* "Genopedia: NOS3," *Center for Disease Control and Prevention*, accessed April 2017, https://phgkb.cdc.gov/HuGENavigator/huGEPedia.do?firstQuery=NOS3&geneID=4846&typeSubmit=GO&check=y&typeOption=gene&which=2&pubOrderType=pubD; "NOS3," *Mala Cards*, accessed April 2017, http://www.malacards.org/search/results/NOS3.

172 *Erectile dysfunction:* Musicki, B., and Burnett, A. L., "eNOS function and dysfunction in the penis," *Experimental Biology and Medicine,* February 2006, https://www.ncbi.nlm.nih .gov/pubmed/16446491.

173 *Contributor to high blood pressure:* Kirchheimer, S., "Sniffing out high blood pressure risk," *WebMD,* 18 February 2003, http://www.webmd.com/hypertension-high-blood-pressure /news/20030218/sniffing-out-high-blood-pressure-risk#1.

174 *BH4, which your body:* Coopen, A., et al., "Depression and tetrahydrobiopterin: The folate connection," *Journal of Affective Disorders,* March–June 1989, https://www.ncbi .nlm.nih.gov/pubmed/2522108; Liang, L. P., and Kaufman, S. "The regulation of dopamine release from striatum slices by tetrahydrobiopterin and L-arginine-derived nitric oxide," *Brain Research,* 3 August 1998, https://www.ncbi.nlm.nih.gov/pubmed /9685635.

175 *Recurrent miscarriage, congenital birth defects, and preeclampsia:* Leonardo, D. P., et al., "Association of nitric oxide synthase and matrix metalloprotease single nucleotide polymorphisms with preeclampsia and its complications," *PLOS One,* 28 August 2015, https://www.ncbi.nlm.nih.gov/pubmed/26317342.

175 *[Heart disease increases] after menopause:* "Hormone replacement therapy and your heart," *Mayo Clinic,* 09 July 2015, http://www.mayoclinic.org/diseases-conditions/menopause /in-depth/hormone-replacement-therapy/art-20047550.

176 *Cardiovascular risk increases:* Hayashi, T., et al., "Effect of estrogen on isoforms of nitric oxide synthase: Possible mechanism of anti-atherosclerotic effect of estrogen," *Gerontology,* 15 April 2009, http://www.karger.com/Article/Abstract/213883.

176 *[Statins] support NOS3:* Cerda, A., et al., "Role of microRNAs 221/222 on statin induced nitric oxide release in human endothelial cells," *Arquivos Brasileiros de Cardiologia,* March 2015, https://www.ncbi.nlm.nih.gov/pmc/articles/PMC4386847.

176 *Serious side effects [of statins]:* "Side effects of cholesterol-lowering statin drugs," *WebMD,* accessed April 2017, http://www.webmd.com/cholesterol-management/side-effects-of-statin -drugs#1.

176 *Don't seem to work well if your NOS3 is dirty:* Hsu, C. P., et al., "Asymmetric dimethylarginine limits the efficacy of simvastatin activating endothelial nitric oxide synthase," *Journal of the American Heart Association,* 18 April 2016, https://www.ncbi.nlm.nih.gov/pubmed /27091343.

177 *Nitroglycerin resistance:* Münzel, T., et al., "Effects of long-term nitroglycerin treatment on endothelial nitric oxide synthase (NOS III) gene expression, NOS III-mediated superoxide production, and vascular NO bioavailability," *Circulation Research,* 7 January 2000, https://www.ncbi.nlm.nih.gov/pubmed/10625313.

177 *Smokers don't typically have success with nitroglycerin:* Haramaki, N., et al., "Long-term smoking causes nitroglycerin resistance in platelets by depletion of intraplatelet glutathione," *Arteriosclerosis, Thrombosis, and Vascular Biology,* November 2001, https:// www.ncbi.nlm.nih.gov/pubmed/11701477.

177 *If your NOS3 is uncoupled:* Daiber, A., and Münzel, T., "Organic nitrate therapy, nitrate tolerance, and nitrate-induced endothelial dysfunction: Emphasis on redox biology and oxidative stress," *Antioxidants & Redox Signaling,* 10 October 2015, https://www.ncbi .nlm.nih.gov/pubmed/26261901.

178 *"Stealing" it from other genes, including NOS3:* Pernow, J., and Jung, C. "Arginase as a potential target in the treatment of cardiovascular disease: Reversal of arginine steal?" *Cardiovascular Research,* 1 June 2013, https://www.ncbi.nlm.nih.gov/pubmed/23417041.

178 *Bacteria in your microbiome:* Cunin, R., et al., "Biosynthesis and metabolism of arginine in bacteria," *Microbiological Reviews,* September 1986, http://europepmc.org/backend /ptpmcrender.fcgi?accid=PMC373073&blobtype=pdf.

178 *It didn't work:* Giam, B., et al., "Effects of dietary l-arginine on nitric oxide bioavailability in obese normotensive and obese hypertensive subjects," *Nutrients,* 14 June 2016, https://www.ncbi.nlm.nih.gov/pubmed/27314383.

178 *[BH4 to] support NOS3 and nitric oxide production:* Vásquez-Vivar, J., et al., "Altered tetrahydrobiopterin metabolism in atherosclerosis: Implications for use of oxidized tetrahydrobiopterin analogues and thiol antioxidants," *Arteriosclerosis, Thrombosis, and Vascular Biology,* 1 October 2002, https://www.ncbi.nlm.nih.gov/pubmed/12377745.

178 *Others found no benefit:* Mäki-Petäjä, K. M., et al., "Tetrahydrobiopterin supplementation improves endothelial function but does not alter aortic stiffness in patients with rheumatoid arthritis," *Journal of the American Heart Association,* 19 February 2016, https://www .ncbi.nlm.nih.gov/pubmed/26896473.

179 *How your body uses arginine:* Förstermann, U., and Sessa, W. C. "Nitric oxide synthases: regulation and function," *European Heart Journal,* April 2012, https://www.ncbi.nlm.nih .gov/pmc/articles/PMC3345541.

179 *Level of BH4 decreases:* Smith, Desirée E. C., et al., "Folic acid, a double-edged sword? Influence of folic acid on intracellular folate and dihydrofolate reductase activity," *Semantic Scholar,* accessed January 2017, https://pdfs.semanticscholar.org/d934/683 d6176b469ff636c4e202b8f99f6bb7217.pdf.

180 *Sleep apnea [and NOS3]:* Badran, M., et al., "Nitric oxide bioavailability in obstructive sleep apnea: Interplay of asymmetric dimethylarginine and free radicals," *Sleep Disorders,* 2015, https://www.ncbi.nlm.nih.gov/pmc/articles/PMC4438195.

181 *You'll end up with elevated levels of homocysteine:* Selley, M. L., "Increased concentrations of homocysteine and asymmetric dimethylarginine and decreased concentrations of nitric oxide in the plasma of patients with Alzheimer's disease," *Neurobiology of Aging,* November 2003, https://www.ncbi.nlm.nih.gov/pubmed/12928048.

181 *High ADMA levels … including dementia:* Selley, M. L., "Increased concentrations of homocysteine and asymmetric dimethylarginine and decreased concentrations of nitric oxide in the plasma of patients with Alzheimer's disease," *Neurobiology of Aging,* November 2003, https://www.ncbi.nlm.nih.gov/pubmed/12928048.

181 *[Dementia and] heart disease:* Brunnström, H. R., and Englund, E. M. "Cause of death in patients with dementia disorders," *European Journal of Neurology,* April 2009, https:// www.ncbi.nlm.nih.gov/pubmed/19170740.

182 *If oxidative stress is present:* Kirsch, M., et al., "The autoxidation of tetrahydrobiopterin revisited," *Journal of Biological Chemistry,* 24 April 2003, http://www.jbc.org/content/278 /27/24481.abstract; Vásquez-Vivar, J., "Tetrahydrobiopterin, superoxide and vascular dysfunction," *Free Radical Biology and Medicine,* 21 July 2009, https://www.ncbi.nlm.nih .gov/pmc/articles/PMC2852262.

11: PEMT: Cell Membrane and Liver Problems

190 *In fact, without a membrane:* Reisfeld, R. A., and Inman, F. P., eds., *Contemporary Topics in Molecular Immunology* (New York: Springer, 2013), 173.

190 *[Phosphatidylcholine] for several important roles:* Vance, D. E., Li, Z., and Jacobs, R. L., "Hepatic phosphatidylethanol-amine n-methyltransferase, unexpected roles in animal biochemistry and physiology," *The Journal of Biological Chemistry*, 16 November 2007, http://www.jbc.org/content/282/46/33237.full.pdf; "Choline," *Oregon State University Linus Pauling Institute's Micronutrient Information Center*, accessed April 2017, http://lpi.oregonstate.edu/mic/other-nutrients/choline.

191 *Helps package and move triglycerides:* "Choline," *Oregon State University Linus Pauling Institute's Micronutrient Information Center*, accessed April 2017, http://lpi.oregonstate.edu/mic/other-nutrients/choline.

193 *That could be a cause of cancer:* Gerl, R., and Vaux, D., "Apoptosis in the development and treatment of cancer," *Carcinogenisis*, February 2005, https://academic.oup.com/carcin/article/26/2/263/2476038/Apoptosis-in-the-development-and-treatment-of.

195 *The higher her risk of breast cancer:* Zeisel, S. H., and da Costa, K. A., "Choline: An essential nutrient for public health," *Nutrition Reviews*, November 2009, https://www.ncbi.nlm.nih.gov/pmc/articles/PMC2782876.

199 *Dirty PEMT contribute to fatty liver:* Song, J., et al., "Polymorphism of the PEMT gene and susceptibility to nonalcoholic fatty liver disease (NAFLD)," *FASEB Journal*, August 2005, https://www.ncbi.nlm.nih.gov/pubmed/16051693.

200 *Neural tube defects, such as spina bifida:* Shaw, G. M., et al., "Choline and risk of neural tube defects in a folate-fortified population," *Epidemiology*, September 2009, https://www.ncbi.nlm.nih.gov/pubmed/19593156.

200 *Decreased memory and more learning disabilities:* Boeke, C. E., et al., "Choline intake during pregnancy and child cognition at age 7 years," *American Journal of Epidemiology*, 15 June 2013, https://www.ncbi.nlm.nih.gov/pmc/articles/PMC3676149; Wu, B. T., et al., "Early second trimester maternal plasma choline and betaine are related to measures of early cognitive development in term infants," *PLOS One*, 2012, https://www.ncbi.nlm.nih.gov/pubmed/22916264.

200 *Most pregnant women in the United States are choline-deficient:* Zeisel, S. H., and da Costa, K. A., "Choline: an essential nutrient for public health, *Nutrition Reviews*, November 2009. https://www.ncbi.nlm.nih.gov/pubmed/19906248; Zeisel, S. H., "Choline: Critical role during fetal development and dietary requirements in adults," *Annual Review of Nutrition*, 2006, https://www.ncbi.nlm.nih.gov/pmc/articles/PMC2441939.

12. Soak and Scrub: Your First Two Weeks

211 *Organically grown foods … more nutritional content:* Aubrey, A., "Is organic more nutritious? New study adds to the evidence," *NPR*, 18 February 2016, http://www.npr.org/sections/thesalt/2016/02/18/467136329/is-organic-more-nutritious-new-study-adds-to-the-evidence.

211 *Fruits and vegetables are those to avoid if not buying organic:* "All 48 fruits and vegetables

with pesticide residue data," *Environmental Working Group (EWG)*, accessed April 2017, https://www.ewg.org/foodnews/list.php.

15. Spot Cleaning: Your Second Two Weeks

284 *[Metformin] increasing histamine:* Yee, S. W., et al., "Prediction and validation of enzyme and transporter off-targets for metformin," *Journal of Pharmacokinetics and Pharmacodynamics*, October 2015, https://www.ncbi.nlm.nih.gov/pubmed/26335661.

284 *[Aspirin and other NSAIDs] ... increased histamine release:* Matsuao, H., et al., "Aspirin augments IgE-mediated histamine release from human peripheral basophils via Syk kinase activation," *Allergology International*, December 2013, https://www.ncbi.nlm.nih.gov /pubmed/24153330; Pham, D. L., et al., "What we know about nonsteroidal anti-inflammatory drug hypersensitivity," *Korean Journal of Internal Medicine*, 5 March 2016, https:// www.ncbi.nlm.nih.gov/pmc/articles/PMC4855107/pdf/kjim-2016–085.pdf.

286 *[Sweating] helps your body expel:* Genius, S. J., et al., "Blood, urine, and sweat (BUS) study: Monitoring and elimination of bioaccumulated toxic elements," *Archives of Environmental Contamination and Toxicology*, August 2011, https://www.ncbi.nlm.nih .gov/pubmed/21057782.

286 *[Damaged glutathione] can contribute to further cell damage:* Mulherin, D. M., Thurnham, D. I., and Situnayake, R. D., "Glutathione reductase activity, riboflavin status, and disease activity in rheumatoid arthritis," November 1996, https://www.ncbi.nlm.nih .gov/pubmed/8976642; Taniguchi, M., and Hara, T., "Effects of riboflavin and selenium deficiencies on glutathione and its relating enzyme activities with respect to lipid peroxide content of rat livers," *Journal of Nutritional Science and Vitaminology*, June 1983, https:// www.ncbi.nlm.nih.gov/pubmed/6619991.

290 *[Oral estrogen] can lead to hypothyroidism:* Mazer, N. A., "Interaction of estrogen therapy and thyroid hormone replacement in postmenopausal women," *Thyroid: Official Journal of the American Thyroid Association*, 2004, https://www.ncbi.nlm.nih.gov/pubmed/15142374.

299 *Further lower your blood pressure:* Bays, H. E., and Rader, D. J., "Does nicotinic acid (niacin) lower blood pressure?" *International Journal of Clinical Practice*, January 2009, https://www.ncbi.nlm.nih.gov/pmc/articles/PMC2705821.

299 *Sauna ... stimulating your NOS3:* Sobajima, M., et al., "Repeated sauna therapy attenuates ventricular remodeling after myocardial infarction in rats by increasing coronary vascularity of noninfarcted myocardium," *The American Journal of Physiology-Heart and Circulatory Physiology*, August 2011, https://www.ncbi.nlm.nih.gov/pubmed/21622828.

Appendix A: Lab Testing

311 *Holotranscobalamin:* "Vitamin B$_{12}$, active; holotranscobalamin," *Dr. Lal PathLabs*, accessed April 2017, https://www.lalpathlabs.com/pathology-test/vitamin-b12-active -holotranscobalamin.

312 *Intrinsic factor test:* "Intrinsic factor blocking antibody," *Specialty Labs*, accessed April 2017, http://www.specialtylabs.com/tests/details.asp?id=568.

316 *Riboflavin deficiency:* "Organix profile interpretive guide," *Genova Diagnostics*, 2014, https://www.gdx.net/core/interpretive-guides/Organix-IG.pdf.

318 *At-home sleep testing:* "Home sleep test and sleep apnea sleep study testing," *American Sleep Association,* accessed April 2017, https://www.sleepassociation.org/home-sleep-test-sleep-apnea-testing.

318 *NovaSom, for home test kits:* "AccuSom at home sleep testing," *NovaSom,* accessed April 2017, http://www.novasom.com.

318 *[Low DHEA-S and] muscle weakness:* Stenholm, S., et al., "Anabolic and catabolic biomarkers as predictors of muscle strength decline: The InCHIANTI study," *Rejuvenation Research,* February 2010, https://www.ncbi.nlm.nih.gov/pmc/articles /PMC2883504.

318 *[Elevated ALT and] phosphatidylcholine levels:* Vance, D. E., "Phospholipid methylation in mammals: From biochemistry to physiological function," *Biochimica et Biophysica Acta.,* June 2014, https://www.ncbi.nlm.nih.gov/pubmed/24184426.

319 *[High TMAO and] kidney issues:* Mueller, D. M., et al., "Plasma levels of trimethylamine-N-oxide are confounded by impaired kidney function and poor metabolic control," *Atherosclerosis,* December 2015, https://www.ncbi.nlm.nih.gov/pubmed/26554714.

319 *[GGT as] early marker of fatty liver:* Bayard, M., Holt, J., and Boroughs, E., "Nonalcoholic fatty liver disease," *American Family Physician,* 1 June 2006, http://www.aafp.org/afp/2006/0601 /p1961.html.

319 *[Fatty liver calculator as] aid for you and your health professional:* "Fatty liver index (FLI) of Bedogni et al for predicting hepatic steatosis," *Medical Algorithms Company,* accessed April 2017, https://www.medicalalgorithms.com/fatty-liver-index-fli-of-bedogni-et-al -for-predicting-hepatic-steatosis.

319 *[Choline deficiency and] reduced blood concentrations of LDL cholesterol:* "Choline," *Oregon State University Linus Pauling Institute's Micronutrient Information Center,* accessed April 2017, http://lpi.oregonstate.edu/mic/other-nutrients/choline.

Appendix C: Mold and Indoor Air Quality Testing

328 *Problems and solutions for your indoor air:* "How to know if your air is unhealthy," *American Lung Association,* accessed April 2017, http://www.lung.org/our-initiatives/healthy-air /indoor/at-home/how-to-know-if-your-air-is-unhealthy.html.

Resources

In this Resources section, I provide the name and brief description of a wide range of products and services that can help you live the Clean Genes life—products and services relating to the air you breathe and the water you drink, as well as to breathing, health professionals, your house and yard, and personal-care products (including supplements).

Air and Water

Clean air and clean water are absolute musts. No exceptions. If either of these is dirty, your genes are dirty. The following products are what I use in my own home. The results are fantastic, and your health will be too. No compromises.

Air

- **Alen Air Purifiers.** Quality air purifiers that also look beautiful and are compact. https://www.alencorp.com

- **Alen Air Dehumidifiers.** Well-made dehumidifiers that will keep your air dry, and thus less likely to be full of mold and dust mites. https://www.alencorp.com

- **Essential oils.** Highly concentrated plant oils that nurture physical and emotional well-being. There are many types of essential oils to choose from. You have to be careful of the source, though: you're looking for organic oils, ideally produced not via solvent extraction but by steam distillation or fractional distillation. Keep essential oils out of the reach of children, because they can be toxic if improperly used. A great resource is the National Association for Holistic Aromatherapy. https://naha.org

Water

- **10 Stage Countertop Water Filter by New Wave Enviro.** A water filter that does a great job and is inexpensive. This was my first water filter, and we still use it when we travel. (Yes, we take a water filter with us, along with a wrench, so that we can tap in to the hotel or timeshare sink and get filtered water.) https://www.newwaveenviro.com

- **Akai Ionizer from High Tech Health.** An option that lets you keep the acidic water for cleaning and watering your plants while you drink the alkaline filtered water. We used this device for fourteen years. Mention "Dr. Lynch" to receive a discount. (*Disclosure:* For this one, I do receive a commission.) http://hightechhealth.com

- **Berkey Water Filter.** A quality water filter that removes many compounds, including fluoride (if you choose to get that add-on). http://www.berkeyfilters.com

- **Premium Shower Filter by New Wave Enviro.** A product we've installed on all our showerheads at home to filter the chlorine out. Try this, and your skin and lungs will thank you. https://www.newwaveenviro.com

- **Rainshow'r Bath Ball.** A product that I used for a long time but eventually stopped using because, although it does remove the chlorine, it's big, bulky, and develops mold. However, if you replace it often, and let it dry out after the bath by hanging it, you should be good to go.

Breathing

It's not easy. Breathing is a skill that we take for granted, yet most of us do it absolutely wrong. Below are resources to help you become a breathing master:

- **Neti Pot.** A device to help rinse mucous from your sinuses and allow air passage through your nose. Ideally, use warm filtered water and a pinch of sea salt. I like to use this in the shower, as it's easiest there, and morning—my shower time—is when I'm most often congested. It can be used over the sink as well.

- **Xlear Sinus Spray.** A big help when your nose is congested. Use a spray or two of this in each nostril (perhaps a few times a day) to help break up the mucous so that it can be blown out. http://www.xlear.com

- **Pranayama breathing techniques.** Taught in many yoga courses, likely near you. In addition, you can find some exercises posted online, thanks to *Yoga Journal.* http://www.yogajournal.com/category/poses/types/pranayama

- **Buteyko.** A Russian breathing method that's used to treat asthma, anxiety, and other conditions. Learn more about how Buteyko might benefit you. http://www.buteyko.com

- **NeuralCranial Restructuring.** A technique that's useful to correct many deviated septums and restore proper breathing. http://www.ncrdoctors.com

Food

- **Thrive Market.** Organic healthy food delivered to your door. https://www.thrivemarket.com

Health Professionals

It's not easy to find a health professional who understands nutrition and biochemistry while viewing medicine through an integrative mindset.

The organizations listed below are ones that I trust. Except for the individuals listed on my own website—which appears first—I don't know all the health professionals in every directory, but I do know that they think holistically.

- **DrBenLynch.com.** A good source for health professionals trained by me in how best to treat dirty genes. www.drbenlynch.com

- **American College for Advancement in Medicine (ACAM).** A group dedicated to bridging the gap between conventional medicine and complementary or alternative medicine. www.acam.org

- **American Academy of Environmental Medicine** and **Naturopathic Academy of Environmental Medicine.** Health professionals who specialize in how to remove mold, industrial chemicals, or heavy metals and are expert in understanding allergic and sensitive reactions to environmental conditions. www.aaemonline.org and www.naturopathicenvironment.com

- **Institute for Functional Medicine.** One of the fastest-growing integrative medicine organizations worldwide. www.functionalmedicine.org

- **Medical Academy of Pediatric Needs (MAPS).** An outstanding organization for children with chronic disease or autism. http://www.medmaps.org

- **American Association of Naturopathic Physicians.** Licensed NDs listed nationally or through their local state association websites. www.naturopathic.org; see also state websites such as www.wanp.org and www.calnd.org

- **International Society for Orthomolecular Medicine.** An organization that the famed Linus Pauling and Abram Hoffer were part of. www.orthomolecular.org

House and Yard

Most of the products below I use in or around my own home, with a few extra suggestions from my friend Suzi Swope of the wonderfully useful website GurlGoneGreen.com. We don't use any chemicals at all in our house. Eliminate them in your home too, and the health results will be noticeable.

Fertilizer and Soil Amendment

- **Hendrikus Organics.** The absolute leader in natural soil health and restoration. The difference in your landscape and vegetable garden will be dramatic. https://www.hendrikusorganics.com

Food Storage

- **Stasher.** Silicone bags that are the perfect replacement for reusable (and "dirty") plastic bags. Great for the fridge or freezer, too. https://stasherbag.com

- **Bee's Wrap.** A long-lasting, user-friendly natural product to keep your food fresh without toxic plastic wrap. https://www.beeswrap.com

Housecleaning

- **CitraSolv.** For wiping things down that need extra support. https://www.citrasolv.com

- **E-cloth.** Great cleaning cloths that don't require solvents/chemicals to clean with. https://www.ecloth.com

- **Ecover.** Good, clean dishwashing tablets. us.ecover.com

- **Norwex.** Cleaning cloths that don't require solvents/chemicals to clean with. Especially great for mirrors. https://norwex.com

- **White vinegar.** Available in any grocery store. Just dilute it with water and you're good to go.

Laundry

- **CitraSolv.** Great for spot cleaning. https://www.citrasolv.com

- **Molly's Suds.** A great laundry detergent that comes in scented or unscented. https://mollyssuds.com

- **The Simply Co.** A great laundry detergent powder. https://thesimply co.com

- **Wool dryer balls.** Use in place of smelly and wasteful antistatic dryer sheets. You can make your own; they're also available from many vendors.

Weeds

- **Propane torch.** A classic tool for open areas that need weeding, such as driveways and walkways.
- **"Hula" hoe.** The best type of hoe for fast removal of weeds.
- **Pitchfork.** The best way to weed a bed. Stick the pitchfork in the soil, tilt back, and remove the fork. Do this for the entire bed. Then bend down and easily pull the weeds out like butter.

Personal-Care Products

The chemicals found in typical personal-care products aren't suitable for you or your genes—so which products can you use instead? As I noted above, my friend Suzi of the highly recommended GurlGoneGreen.com has kindly put together a list of her favorite clean-green products for me so that you can have some gene-friendly options.

Bath Products

- **Acure Body Wash for Kids.** A body wash that makes a fantastic bubble bath. https://www.acureorganics.com

Cosmetics

- **100% PURE.** A huge line offering everything from makeup to skin care. Their lengthening mascara is a great option. https://www.100 percentpure.com
- **Crunchi.** A great clean makeup line. Love their mascara, primer, foundation, and blush. https://crunchi.com
- **Dusty Girls.** Love their bronzer, blushes, and BB cream—all budget-friendly. (BB cream is a type of foundation that's not as heavy as normal liquid foundation but not as light as tinted moisturizer.) http://dustygirls.com

- **GIA Minerals.** Love their mascara and eyeshadow selection. https://www.giaminerals.com

- **Hynt Beauty.** Great all-around clean makeup line. Love their eyebrow cream, mascara, and concealer. https://www.hyntbeauty.com

- **Ilia Beauty.** Great lipsticks and lip crayons. https://iliabeauty.com

- **Kjaer Weis.** Great foundation, blushes, and lip products. https://kjaerweis.com

- **Lily Lolo.** A great all-around makeup line. Love their mascara, eyeshadow palettes, eyeliner, and BB cream. https://www.lilylolo.us

- **RMS Beauty.** One of the first natural makeup lines. Great easy-to-wear products, especially their Un Cover-up, Lip2Cheeks, and powders. https://www.rmsbeauty.com

- **Root Pretty.** Love their Pearl Powder Mineral Foundation. Budget-friendly. https://www.rootpretty.com

- **Vapour Organic Beauty.** An amazing line, including foundations, blushes, and lip products. www.vapourbeauty.com

- **W3ll People.** A good clean makeup line. w3llpeople.com

- **Zuzu Luxe.** Love their liquid liners. Budget-friendly and available at Whole Foods. https://gabrielcosmeticsinc.com/brand/zuzu-luxe

Deodorants

- **Green Tidings.** Great, effective natural deodorant. http://www.greentidings.org

- **Primally Pure.** Effective and budget-friendly. Sensitive formula available. https://primallypure.com

- **Rustic Maka Pachy.** Comes in a variety of natural scents and is budget-friendly. https://www.rusticmaka.com

- **Schmidt's.** Effective and budget-friendly; sensitive formula available. https://schmidtsnaturals.com

- **Ursa Major Hoppin' Fresh Deodorant.** Effective; comes in unisex scent. https://www.ursamajorvt.com

Hair Products

- **Acure.** Budget-friendly shampoos, conditioners, and styling products. Can be found at Whole Foods, health stores, and online. https://www.acureorganics.com

- **Flourish Organic Hair.** A variety of hair-care products from shampoos and conditioners to styling products. Budget-friendly. www.flourishbodycare.com

- **Green & Gorgeous Dry Shampoo.** Available in options for light and dark hair. https://gandgorganics.com

- **Hairprint.** A very clean hair dye made from food-grade ingredients. https://www.myhairprint.com

- **Herbivore Sea Mist Spray.** Hair texturizer and sea salt spray. https://www.herbivorebotanicals.com

- **Innersense Organic Beauty.** Natural salon-quality shampoos, conditioners, and styling products. Their shampoo and conditioner are especially great for color-treated hair. https://innersensebeauty.com

- **Josh Rosebrook.** The best natural hair spray and the best spray volumizer ("Lift"). Created by a stylist. https://joshrosebrook.com

- **Primally Pure Dry Shampoo.** A great nontoxic dry shampoo. https://primallypure.com

- **Rahua.** Great volumizing shampoo and conditioner. https://rahua .com/us

- **True Botanicals.** Ultra-luxe clean shampoo and conditioner in eco-friendly pump bottles. https://truebotanicals.com

Hand Creams

- **100% Pure Hand Buttercream.** Comes in a tube and moisturizes with no greasy residue. https://www.100percentpure.com

- **Osmia Organics Vanilla Shea Hand Cream.** A great hand cream that doesn't leave behind a residue. Smells amazing. https://osmia organics.com

- **Shea Terra Organics Mini Shea Whippers.** Come in cute jars and are wonderfully hydrating and clean. https://www.sheaterra organics.com

- **Zoe Organics Everything Balm.** Great for hands or anywhere you need some extra hydration. https://www.zoeorganics.com

Hand Soaps

- **Kosmatology.** Best hand soap out there! www.kosmatology.com

Lip Balms

- **Henné Organics.** Great lip balm—really smooth formula—and great lip exfoliant. https://henneorganics.com

- **Hurraw Lip Balm.** Great lip balm packaged in tubes; available in a variety of flavors. https://hurrawbalm.com

- **Kari Gran Lip Whip.** Ultimate lip moisturizer; comes in a jar. https://karigran.com

Perfumes

- **Florescent.** Luxury perfumes that come in spritzer form. https://florescent.co

- **Josh Rosebrook Ethereal Botanical Fragrance.** A truly luxurious natural fragrance. Makes switching to a nontoxic fragrance easy. https://joshrosebrook.com

- **Lotus Wei.** Plant-based scents that come in a variety of application methods, from roller balls to spritzers. Products help with moods. https://www.lotuswei.com

- **LURK.** A ton of natural scents to choose from. https://lurk made.com

Skin-Care Products

- **Acure.** Body and facial skin-care products that are budget-friendly. https://www.acureorganics.com

- **Dr. Bronner's.** Great castile soaps for use as body washes; DIY cleaning recipes also available. https://www.drbronner.com

- **Josh Rosebrook.** A line based on herbs and plant power. https://joshrosebrook.com

- **Kahina Giving Beauty.** A skin-care line based on argan oil. kahina-givingbeauty.com

- **Kosmatology.** Great body washes, facial products, and scrubs. Budget-friendly. www.kosmatology.com

- **Leahlani Skincare.** This skin-care line is not only effective and clean, but offers something for every skin type. Also budget-friendly. https://www.leahlaniskincare.com

- **Laurel Whole Plant Organics.** A skin-care line based on herbs and flowers that's 100 percent raw, organic, and unrefined. https://www.laurelskin.com

- **Live Inspired Organics.** Best scrub and body butter. www.liveinspiredorganics.com

- **Marie Veronique.** A skin-care line based on science and research. It offers something for every skin type. https://www.marieveronique.com

- **May Lindstrom.** Great facial masks. https://maylindstrom.com

- **Maya Chia Beauty.** A superb antiaging line that uses chia seed oil as the base for all formulations. https://mayachia.com

- **Osmia Organics.** Best bar soaps. Great body oils as well, and facial products. https://osmiaorganics.com

- **True Botanicals.** This skin-care line has studies backing its potency and results. Great for those who suffer from acne at any age, and a great antiaging line. https://truebotanicals.com

Sunscreens

- **Babo Botanicals SPF 40 Daily Sheer Facial Sunscreen.** Sheer and lightweight. www.babobotanicals.com

- **DeVita Solar Body Moisturizer Mineral Sunscreen SPF 30+.**
  Effective and budget-friendly. www.davita.com

- **Loving Naturals Adorable Baby Sunscreen SPF 30+.** Great for
  babies and kids, with simple ingredients. https://lovingnaturals.com

- **Suntegrity Mineral Sunscreen.** Sunscreen and self-tanners.
  www.suntegrityskincare.com

- **Raw Elements Sunscreen.** The cleanest sunscreen out there.
  Comes in a variety of formulas, from stick to tube to tin form.
  https://rawelementsusa.com

Toothpastes

- **Jason's Toothpaste.** The one we use in our house. Love it! http://
  www.jason-personalcare.com

- **Tom's of Maine Toothpaste.** Also a family favorite. http://www
  .tomsofmaine.com/home

- **Uncle Harry's Toothpaste.** Easy to find at Whole Foods; leaves you
  with a fresh, clean feeling. www.uncleharrys.com

- **Wellness Mama Blog.** Great DIY recipes for toothpastes that will
  restore minerals back into your teeth—an approach that some people have used (along with a careful diet) to prevent cavities. https://
  wellnessmama.com

Saunas

A sauna is an absolute must-purchase item unless you have access to a
gym with a great sauna. The saunas below are all of great quality, though
they look different and the heaters are a bit different. I've negotiated significant discounts for you. I do receive a commission for recommending
these saunas.

- **Sunlighten Saunas.** Beautifully designed, low electromagnetic frequency (EMF), and built with high-quality, safe materials. Heaters
  are full spectrum—near-, far-, and midrange. Sunlighten also has a
  one-person sauna just big enough for you to lie down in. Mention "Dr.
  Lynch" to receive a significant discount. http://www.sunlighten.com

- **HighTech Health.** Constructed with high-quality, safe materials and low electromagnetic frequency (EMF); designed for individuals with multiple chemical sensitivities; easy to set up and move as needed. I owned a three-person HighTech sauna for ten years. HighTech offers very low-EMF-emitting heaters, good air circulation, light therapy, and a setup for music. Mention "Dr. Lynch" to receive significant savings. http://hightechhealth.com

Supplements

As you take supplements, always keep the Pulse Method in mind. (See chapter 12 for a refresher.) Below are listed supplements I've formulated as well as brands I recommend based on how well they've performed for my clients and fellow health professionals.

- **Bio-Botanical Research** (https://biocidin.com)
  —**Liposomal Biocidin.** An effective antimicrobial.
  —**Biocidin Throat Spray.** Fantastic to support a sore throat.

- **Seacure** (http://www.seacure-protein.com)
  —**Hydrolyzed White Fish.** Excellent for healing the gut.

- **Seeking Health** (my company: www.seekinghealth.com)
  —**5-HTP.** Supports serotonin production.
  —**Adrenal Cortex.** Supports adrenals to help you wake up in the morning.
  —**DIM + I3C.** A blend that supports healthy estrogen levels.
  —**HistaminX.** To help counter symptoms of histamine excess.
  —**HomocysteX Plus.** Supports the Methylation Cycle and healthy homocysteine levels.
  —**Lithium.** Supports serotonin levels and promotes a sense of calm.
  —**Molybdenum.** Helps process sulfites. We offer it in liquid 25 microgram doses and 75 microgram capsules.
  —**Multivitamins.** Many to choose from—powders, chewables, capsules.
  —**NADH + CoQ10.** Helps get you out of bed in the morning and eliminates the need for caffeine if you're trying to quit.

—**Neutralize.** Helps reduce a mild histamine reaction or "off" feelings from an environmental exposure.

—**Optimal Electrolyte.** A complete electrolyte formula without sugar.

—**Optimal GI Powder.** A comprehensive gut lining repair formula.

—**Optimal Liposomal Glutathione Plus**: Provides glutathione right to your cells, along with the nutrients needed to use and recycle glutathione.

—**ProBiota Bifido.** Helps break down histamine from bacteria, food, and drink.

—**ProBiota HistaminX.** Helps break down histamine from food, drink, and bacteria. Helps solve the problem of high histamine by improving your microbiome.

—**Pro-Digestion Intensive.** A comprehensive digestive enzyme.

—**PreGestion.** Provides stomach acid to help reduce belching and aid digestion.

—**Optimal Adrenal.** A nonstimulating, adaptogenic blend of herbs and nutrients to take the edge off stress without sleepiness. A favorite for parents.

—**Optimal Prenatal Protein Powder.** My favorite product by far: a complete prenatal with protein powder that can be drunk as a smoothie. This is what I drink most mornings for my breakfast, because it's fast and complete—although I do not intend to get pregnant! This is what I use in the smoothie recipes on pages 237 and 243.

—**Optimal Detox Powder.** A comprehensive detoxification formula with protein powder; makes a great breakfast shake. This is another protein powder blend that I use in my smoothie recipes.

—**Optimal Start.** A great foundational multivitamin for those who are sensitive to methyl donors; contains no folate or B_{12}, compounds that some people are sensitive to.

—**Optimal Sleep.** A combination of nutrients to help people relax and have a good night's sleep without feeling hungover the next day.

—**Ox Bile.** Supports fat digestion, supports those without a gall-bladder, and acts as an antimicrobial in the small intestine to help combat SIBO.

—**PQQ.** Useful for those experiencing oxidative stress or significant postworkout soreness.

- **US Enzymes** (https://usenzymes.com)

—**Digestxym.** A plant-based comprehensive digestive enzyme.

—**Serraxym.** A systemic enzyme to help break down cellular debris.

Tracking Products for Exercise, Food, Heart-Rate Variability (HRV), and Sleep

Understanding how your body behaves in real time is extremely useful. It's also useful to know exactly what you're eating, how much, and what that food contains. These tracking products can help you gather and interpret such data.

- **CRON-O-Meter app.** Track what you eat and how much of it. This app shows you in real time how much more you should eat for the day, broken down into protein, fats, and carbs. It also has settings for various diets, such as Paleo or Ketogenic. If you want to lose or gain weight, it adjusts for that as well. I learned a lot about my eating by using this app. Available on the app store of your phone. https://cronometer.com

- **HRV4Training.** A great phone app that tracks your HRV via your cellphone camera. http://www.hrv4training.com

- **Nutrient Optimiser.** Ties in with CRON-O-Meter to inform you what you should eat, what you shouldn't eat, and what you should eat more of—with the goal of getting all your nutrients from food. A fantastic program designed by Marty Kendall. https://nutrientoptimiser.com

- **ŌURA ring.** A hard-working and effective gadget. I had no idea how bad my sleep was until I started tracking it. ŌURA ring woke me up to the fact I wasn't sleeping well. I made changes, and now my sleep is fantastic (unless I work late or eat late!). This app also tracks your readiness to exercise (how hard you can exercise, or if you need to take it easy for the day based on HRV, heart rate, and body temperature) and your actual exercise (how much, intensity, when). It's a good-looking and durable ring as well. I get a lot of compliments! Use code "aejjxo2" at ouraring.com to save 10 percent.

- **Sleep Cycle app.** A great starting app to measure the quality of your sleep. It gives you some insight, but the ŌURA ring is more precise. Available on the app store of your phone. https://www .sleepcycle.com

- **WISE. What should I eat?** Tired of people telling you what you can't eat? Answer a few questions about what foods give you symptoms, and about your health more generally, and receive a report in minutes about what you *can* eat. Expand your report into customized recipes prepared by a professional chef! Try WISE (What I Should Eat). www.drbenlynch.com

Index